Pest Control

D0309974

New Studies in Biology

Pest Control

Second Edition

H.F. van Emden

Professor of Applied Entomology,
Departments of Horticulture and Pure and Applied Zoology,
The University of Reading

CAMBRIDGE
UNIVERSITY PRESS

Published by the Press Syndicate of the University of Cambridge
The Pitt Building, Trumpington Street, Cambridge CB2 1RP
40 West 20th Street, New York, NY 10011–4211, USA
10 Stamford Road, Oakleigh, Victoria 3166, Australia

First published in Great Britain by Edward Arnold 1972 as *Pest
Control and its ecology*
Reprinted 1976, 1977, 1980, 1984
Second edition 1989
First published by Cambridge University Press 1992

Printed in Great Britain at the University Press, Cambridge

British Library cataloguing in publication data
Van Emden, H.F. (Helmut F.)
  Pest control. – 2nd ed
  1. Pests control. – 2nd ed
  I. Title   II. Series
  632′.9

Library of Congress cataloguing in publication data available

ISBN 0 521 42788 6 paperback

General Preface to the Series

Charged by its Royal Charter to promote biology and its understanding, the Institute of Biology recognises that it is not possible for any one text book to cover the entirety of a course. If evidence was needed, the success of the *Studies in Biology* series was a testimony to the need for specialist, up-to-date publications in education. The Institute is therefore pleased to collaborate with Cambridge University Press in producing the *New Studies in Biology* series.

The new series is set to provide as great a boon to the new generation of students as the original did to their parents.

Suggestions and comments from readers will always be welcomed and should be addressed either to the New Studies in Biology Editorial Board at Cambridge University Press or c/o The Publications Division at the Institute.

Robert Priestley
The General Secretary

The Institute of Biology
20 Queensberry Place
London SW7 2DZ

Preface to the First Edition

This slim book cannot attempt to cover pest control in its full diversity; it is limited to the clearly definable area of the control of insect pests on growing crops. There is more to such control today than applying poisonous chemicals. The questions that have to be asked are largely biological and ecological, and their answer affects our food supply, our standard of living and our environment. The text suggests many studies that can be attempted around any farm or in a simple laboratory or glasshouse.

This is not really a book about ecology; it is firmly about pest control. It tries to assess where we are going, and the ecological merits, record to date and future potential of the many approaches. Such an assessment has strong subjective overtones, and another author might well make a contrasting judgement.

Pest control is indeed a 'Study in Biology'. Yet the biologist, and particularly the ecologist, who is going to play an increasing role in the subject in the future, will have maximum impact if he can visualize the total environment of crops to include not only soil, weather, insects and plants but also the social, economic and industrial environments with which the crops are inexorably enmeshed.

Reading, 1974   H. v. E.

Preface to the Second Edition

The new edition remains unchanged from the first edition in respect of the area it covers – the control of insect pests on growing crops. Reviewers of the first edition were kind enough to suggest that its coverage was adequate to merit the title 'Pest Control' without the added slant 'and its Ecology'; their suggestion has therefore been accepted for the title of the new edition. To satisfy the author, however, the first chapter on chemical control has been somewhat enlarged to improve the balance of the book with the acceptance of the new title.

With the enlarged format of the new series, the opportunity has been taken to create a largely new text for several of the chapters and to alter the emphasis given to the various methods of pest control to reflect modern attitudes. Insect pathogens and pheromones have been allocated their own chapters, since their current and future usefulness has become much clearer with time. Pest management, a nebulous concept in 1974, is perhaps now merely shrouded in mist by comparison, and it was felt a more extended discussion was warranted. Activity in connection with plant resistance to pests has also grown enormously in the last twenty years, and considerably more emphasis has been given in this edition to the development and use of this method.

Two forces which have played a major role since 1974 in shaping the pest control of today have been a) the rapid development of resistance to pesticides in many major pests, including to the synthetic pyrethroid insecticides at a time when few new alternative compounds are available and b) ever-increasing public concern (much of it fuelled by unreasonable propaganda) about the hazards insecticides pose to our own health and that of our environment. Certainly it is appropriate to seek ways of reducing the quantities of pesticide used. Major ways of achieving this are by encouraging the use of less wasteful equipment and making more use of restraints on pests other than pesticides; the latter approach is often called 'pest management'. However, many proponents of pest management (including the author) see one of the most important roles of the concept as preserving the useful life of the current, largely safe, excellent and above all flexible arsenal of insecticides. Pest management does not necessarily do away with the need for insecticides; indeed, most *successful* pest management involves their continued use to a reduced extent. If we lose (because of resistance) the flexibility with which our current pesticides can be used to obtain some selectivity between pests and natural enemies, we may also lose the potential to solve our problems with the ecologically sounder approach to pest control that pest management still offers us.

Helmut van Emden
1989

Contents

1

Introduction to chemical control

The history of insecticide use dates back many centuries, certainly to before 1000 BC, when it is mentioned by Homer, but the real landmark in terms of modern agriculture is the spread of the Colorado beetle (*Leptinotarsa decemlineata*) across the United States in the second half of the nineteenth century (see Section 2.4). Food production and the national economy were both threatened by this potato pest and, after much argument, it was finally decided to take the unprecedented step of spraying the potato crops with a human poison (arsenic in the form of Paris Green). The mass human mortality predicted by the prophets of gloom did not occur, and there is no doubt that control of the Colorado beetle with Paris Green opened the way to a widespread use of biocides (destroyers of life in general) on crops destined for human consumption.

1.1 Kinds of insecticides

Until recently, it was possible to include nearly all insecticides under a few groupings, but random testing by industry has now identified insecticidal activity in a large number of compounds outside major chemical groupings. None the less, the major groupings still represent the majority of insecticides in current or past use and form a useful basis for an account of insecticides and their variety in terms of routes to the insect target.

1.1.1 Early insecticides

These were mainly of three kinds. Among the earliest insecticides were toxic extracts of plants long used by primitive tribes to tip their hunting arrows or to bring fish to the surface of rivers and lakes. Best known of these substances are pyrethrum (from a chrysanthemum-like plant), rotenone (a root extract of the derris plant) and nicotine (from tobacco). These plant extracts work in a variety of ways, poisoning either the nervous or respiratory system. They penetrate the cuticle of insects (**contact** insecticides) and are very short-lived (hours or days). The insect therefore has to be contacted by drops of spray (ephemeral contact insecticides) or, in the case of nicotine when burnt, a toxic smoke (**fumigant** insecticide) is inhaled by the insect. The short life of these compounds was initially seen as a disadvantage, but today this 'disadvantage' gives them a special role when crops need treating close to harvest. Many other plants contain toxic chemicals (several are known to be very toxic to man, e.g. hemlock and atropine), and so-called 'natural insecticides' derived from plants can be every bit as deadly as chemicals synthesised by man. However, the word 'natural' is enough to endear plant-derived insecticides to many who are worried about using other insecticides, and the short life of these chemicals after spraying certainly imparts safety to the environment. There is considerable interest shared by industry in discovering new insecticides in plants; e.g., azadirachtin from the tropical neem tree has been researched as an insecticide since the early 1960s in many countries.

The second group of early insecticides were oils which, owing to their deleterious effects on plants, were mainly used only on dormant leafless plants such as apple trees over winter. Oils cover and suffocate insects and mites, including their eggs. Oils are still used in desperation today when mites, particularly, show tolerance to other pesticides, and then they may even have to be used on leafy annuals in spite of the inevitable damage to the plant.

The third group (including Paris Green) were the **stomach poisons**. These were toxic radicals (e.g. of arsenic or fluorsilicate) formulated as salts of metals (e.g. lead or sodium). Such salts were relatively stable, and plants could be sprayed without damage from the poisons, which have very general activity against life by precipitating protein. The salt must be ingested by a leaf-feeding insect before the free toxin (e.g. arsenic) is released in the gut following hydrolysis of the salt. A stomach poison has a major advantage over a contact poison because it is 'addressed' only to a pest consuming the leaves, and predators can move safely over the deposit. However, the stomach poisons are also rather persistent, and therefore there is risk of ingestion by man. Also, the metals on which the salts were based are undesirable long-term soil contaminants.

1.1.2 Residual contact insecticides

At a time when the short-lived plant extracts were used as contact insecticides, there was considerable interest in the possibilities of longer lasting crop protection with insecticides which had a long **residual contact** life, so that insects walking on the dried spray on the leaf would contact a lethal dose of pesticide. In fact, thinking at that time was that the longer the life of the residue, the better. Thus there was enormous welcome for the first residual contact insecticides synthesised by chemists, the persistent organochlorines. The most famous (or infamous, depending on viewpoint) of these is DDT, first synthesised as a molecule in 1874, but its insecticidal properties were not discovered till the late 1930s. For the first time man had a weapon against the malaria mosquito, the lice infesting him in long battle campaigns and cheap long-term protection for his crops. For crop use, the chemical could not only be sprayed, but often applied much more simply as a coating to the seed before

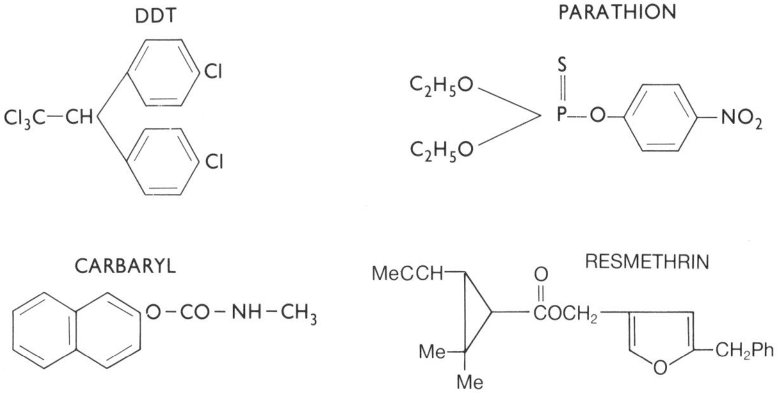

Fig. 1.1 Chemical structure of the four main groups of residual contact insecticides: organochlorine (e.g. DDT); organophosphate (e.g. parathion); carbamate (e.g. carbaryl); synthetic pyrethroid (e.g. resmethrin).

sowing (**seed dressing**) or just harrowed into the soil. DDT and many other organochlorines are not unduly toxic to man, but they break down slowly, and some, like the soil insecticides aldrin and dieldrin, probably remain in the environment for periods approaching the life-span of man. It is the persistence of these compounds, not their toxicity, which has resulted in attempts to phase them out wherever possible. The organochlorines dominated the 1940s and 1950s, and many are still in use today. DDT itself is still widely used in the tropics because of its safety to the farmer applying pesticide, HCH has fumigant properties as well as being a good soil insecticide and endosulfan is a very good general insecticide for the tropics which has been used for many years and has really not created problems.

The mode of action of the organochlorines is multiple and complex; the two most important actions are an inhibition of the enzyme cytochrome oxidase, which mediates gas exchange in the respiration of all animals which use blood as a gas carrier, and a destabilisation of the nervous system.

A second group of residual contact insecticides was produced in the late 1940s – the organophosphates. These are usually highly toxic to man but easily broken down and much less persistent than the organochlorines. Parathion and malathion were among the earliest organophosphates, and the group has been explored thoroughly to produce an arsenal of many diverse, flexible compounds. Many (including malathion) are strongly fumigant, or can move through the leaf from the top to the lower surface (**translaminar action**). Others, especially dimethoate, can 'sink' into the plant tissues to reach stem borers, leaf miners and other concealed insects (**quasi-systemic action**). Several (including metasystox and schradan) are absorbed by the leaves and roots of plants and mobilised upwards and outwards to other parts of the plant. Such **systemic** action can compensate for poor initial coverage with the pesticide and is especially effective against aphids and several other groups which then suck a poisoned sap; yet the plant surface may be quite safe for other insects, including parasites and predators, to walk over.

The organophosphates were increasingly preferred over the organochlorines during the 1960s, and are probably still the most widely used insecticide group today. Their great variety of modes of reaching the insect is their greatest value; no other group of insecticides offers the same flexibility. It has even been possible to replace the organochlorine soil insecticides such as aldrin and dieldrin with organophosphates by incorporating the organophosphates (often diazinon) in a granule on or in the soil. The relatively non-persistent pesticide is continually replaced in the soil as the granule slowly dissolves away. In addition it is possible to surface-coat the granules, creating a 'time-bomb' as far as the initiation of release of pesticide in concerned. Organophosphates are also easily formulated in a shell of inert material as a 'microcapsule' (see Section 9.4.5), resulting in a stomach poison (i.e. toxic only on ingestion and therefore selective for plant-feeding insects).

Like the organochlorines, the organophosphates have a very non-specific mode of action on animals, whether insects or man. They combine with the enzyme cholinesterase, and thus inhibit the hydrolysis of the acetylcholine produced at the nerve endings to carry nerve impulses across the synapses. In poisoned animals, therefore, acetylcholine accumulates at the synapses, giving constant nervous stimulation resulting in tetanic paralysis.

A third group of residual contact insecticides, the carbamates (derivatives of carbamic acid), were introduced in 1956 with the compound carbaryl. The persistence of carbamates lies between that of the organochlorines and organophosphates, as does the toxicity of the first carbamates produced. Carbaryl has been widely used for the control of caterpillars and other surface plant feeders. Methomyl has good contact action, but is also fumigant and slightly systemic. The early carbamates were followed by some highly systemic compounds, some of which (e.g. carbofuran and aldicarb) are very toxic to man. There is also a very unusual systemic and fumigant carbamate, pirimicarb, which has very low mammalian toxicity and a biochemical selectivity for aphids and some flies. When used against aphids, therefore, ladybirds and parasites are not killed.

The action of carbamates, like that of the organophosphates, is on the nervous system by the accumulation of acetylcholine at the nerve synapses. Rather than inhibiting the enzyme cholinesterase, however, they act as competitors with the enzyme for the substrate's surface.

The most recent group of residual contact insecticides are the synthetic pyrethroids. It was long the goal of insecticide chemists to modify natural pyrethrum and impart several additional desirable properties such as photostability to increase persistence. Although partial success came with the synthesis of allethrin as early as 1949, the real success came at Rothamsted Experimental Station in Britain in the early 1970s. The first synthetic pyrethroids combining high toxicity to insects with low mammalian toxicity and greatly increased stability were announced in 1973, and since then many new pyrethroids have been synthesised and marketed. Cypermethrin, for example, is 300 times more toxic than DDT to insects but only 60% as toxic to man. The mode of action of the pyrethroids appears to be a physicochemical process on the nerve membrane sufficiently similar to the action of DDT that insect populations resistant to DDT can show some cross-resistance to pyrethroids.

The similarity to DDT does not end there. Like DDT, the synthetic pyrethroids are purely broad-spectrum residual contact poisons; there is no additional fumigant, translaminar or systemic action. When applied, therefore, these highly potent insecticides are often very damaging to natural enemies, and pest flare-backs resulting from destruction of natural enemies and development of resistance to pyrethroids in the pest have occurred sufficiently often to stimulate second thoughts on exactly how and when these valuable new insecticides should be used.

1.2 Application of pesticides

In brief, pesticides are usually applied in one of three ways:

a) spraying droplets with water, oil or air as the dilutant. The pesticide may be dissolved in the liquid carrier, sprayed as an emulsion or even used as undissolved particles (flowables if the particles are minute or wettable powders if larger);
b) spreading the compound absorbed on or impregnated into an inert solid carrier (dusts and granules);
c) burning the compound to create a pesticidal smoke which will penetrate all parts of a more or less enclosed space (e.g. a dense orchard, a glasshouse).

The subject of pesticide application involves some really fascinating topics such as the fluid kinetics of droplet production; the use of additives (formulation) to impart certain physical characteristics to the spray, such as improved retention on surfaces or better penetration of the insect cuticle; and the engineering aspects of spray outlets (nozzles) and pressure sources (pumps). Much of this, however, lies outside the scope of this series, and readers are referred to Matthews (1979) for an excellent account.

Certainly formulation (the solvent, carrier, additives, etc.) and the method of application (e.g., see Fig. 9.4) can have almost greater influence on the efficiency and selectivity of kill than the choice of active ingredient. How these variables may be manipulated so that the pesticide application is less damaging to natural enemies is discussed in Chapter 9. It is a long, long way in biological terms from the emission of pesticide from a machine to achieving kill of a pest.

The first problem is to get the right amount of chemical onto the target, which is so often the foliage of plants that the problem will be discussed here with reference to that target only. Here the size of the drop, not when it leaves the machine but when it reaches the target, is of critical importance. Unfortunately a spray cloud never consists of identically sized drops. Nearly all spray equipment used on the farm today, as it has always done, relies on forcing

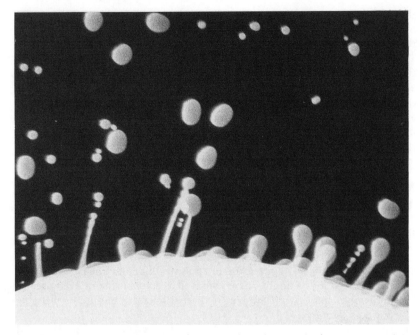

Fig. 1.2 High-speed photograph of spinning disc, showing the fragmentation of the liquid sheet into filaments, which rupture to produce main and satellite drops. Photograph courtesy of ICI Agrochemicals.

liquid through a hole under pressure to produce the spray, and this process (which amounts to disintegrating the edge of an expanding sheet of liquid) produces particularly variable drop sizes. High-speed photography (Fig. 1.2) shows that, at the edge of the sheet of liquid, larger (main) drops are produced on filaments which shatter to produce a larger number of small (satellite) drops. The larger drops contain most of the insecticide a farmer has paid for, but cover very little of the crop surface if they are retained on the foliage at all. It is quite normal for 60% of the total spray volume to be used for only 20% of the drops (the largest ones) formed at the nozzle. These large drops will often bounce off the leaf they contact, particularly if the leaf is hairy or very waxy. This 'spray reflection' may occur with drops about 250 μm in diameter or larger. This reflection, and the effect thereon of different leaf surfaces, can be simply demonstrated by passing leaves or leaf discs laid on filter paper under a source of coloured drops, e.g. a narrow-tipped burette containing a coloured solution, and leaving the tap very slightly open (Fig. 1.3). The other problem with larger drops that are actually retained on the leaf is that they are likely to spread into each other and coalesce; when this happens, the liquid on the leaf collapses into a very thin film, and much runs off the leaf onto the ground.

It is therefore the smaller drops, vast in number but accounting for rather little of the total volume applied, which provide the coverage needed by the

Fig. 1.3 Photographic track of droplets impinging on a reflective leaf surface and 'bouncing' off instead of adhering. Photograph courtesy of Dr. G.D. Dodd.

farmer on the foliage for pest control (Fig. 1.4). However, the smaller drops are much less likely than the larger drops to come in contact with the foliage in the first place. Moreover, small drops have an enormous surface to volume ratio, and will rapidly evaporate and get even smaller as soon as they leave the nozzle, particularly in warm and dry weather. The problem with small drops is that they have very little momentum, and lose speed very rapidly. When we see a cone of spray apparently driving its way into the crop canopy, it is actually the large drops we are seeing; the small ones are mostly invisible to the naked eye. A good way of illustrating the problems of small drops is to crush the end of a piece of chalk, and then to throw different sized fragments at the far wall of a room. It is quite easy to reach the wall with pieces the size of the letters printed on this page, but impossible with the really fine particles. This is because, like the small drops in a spray, they quickly lose any momentum we impart (due to friction with the air), and come to a stop in relation to the air around them. If that air is stationary (it rarely is out of doors), gravity will take over, and the drop will fall under its own weight and 'sediment' onto whatever surface it first encounters. The smaller the drop, the more slowly it will of course fall, and it may well evaporate away to no more than a tiny particle of insecticide before it has made a contact. If the air is moving sideways, it will carry the stationary drop sideways during its fall. As the air approaches an obstacle, e.g. a plant stem or leaf, it will stream around the obstacle, carrying the small drop round

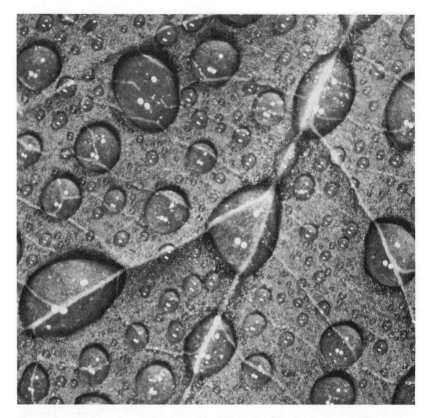

Fig. 1.4   Close-up of part of a sprayed leaf illustrating the distribution of different drop sizes in a spray. The smaller drops provide the bulk of the spray coverage. Photograph courtesy of ICI Agrochemicals.

as well. Even drops which still have a little momentum of their own will tend to follow the air-stream, and may well then accumulate, if at all, on the edges of leaves or on leaves edge-on to the air-flow. Such leaves may collect insecticide deposits seven times greater than leaves facing the spray. All this assumes that the small droplet has got to the crop in the first place. Often in daytime, however, air movement has an upward component resulting from the sun heating the earth; the warm air near the ground then rises to great heights (thermals). That is, of course, also the direction taken by very small drops under those conditions, and insecticide may then 'drift' very long distances, well away from the crop area. Such drift will be most pronounced on a warm still day, when the smoke from chimneys rises vertically upwards. In contrast, small drops sprayed with a mild cross-wind are much more likely to finish in the crop.

In the end, a very small segment of the droplet-size spectrum will both reach the target and be retained thereon. Thus drops in the 30- to 50-μm-diameter may be right for contacting insects resting on foliage, whereas 50- to 100-μm drops are more likely to deposit on and be retained by the foliage itself. Some idea of how small the appropriate segment is in relation to the cost of pesticide to the farmer can be obtained from the results of a test where a spray of Bordeaux mixture (a fungicide based on copper) was applied to cocoa leaves to leave a theoretical deposit of 25 μg of copper per square millimetre of leaf. The maximum deposit achieved was 1.3 μg/mm^2 on the lower surface of the young leaves; on the waxy surface of mature leave the deposit fell to 0.7 μg/mm^2.

A typical amount of pesticide (quoted as amount of active ingredient, which is often a half or less of the volume of the commercial formulation the farmer purchases) applied per hectare is 750 g. The average deposit in the example just quoted would mean a farmer would be lucky to leave 30 g of the 750 g he has paid for on his crop plants; the rest would end as a contaminant on the soil or would be carried away in the air. In practice, the farmer cannot usually match the very carefully controlled application used in the cocoa experiment; a more realistic figure would be that 10 rather than 30 g/ha would be deposited from a 750-g/ha spray.

Is this waste and consequent loss of insecticide to the environment inevitable? Unfortunately at the present time the answer is 'yes', although better maintenance of equipment and choice of conditions for spraying could no doubt cut the wastage to some extent. However, three developments in pesticide application technology hold the potential for future improvement of the situation.

The first has arisen from the need for spray equipment for fixed-wing aircraft to have a high output per minute if an adequate dose per hectare is to be achieved at ground speeds of about 100 to 150 km/h. A typical 100-l/min output requires about 24 traditional nozzles on the aircraft's boom, but the same output can be achieved by four Micronair spinning cages. The Micronair (Fig. 1.5) involves the spraying of pesticide through a diffuser tube onto a metal gauze cage spun rapidly by small propellors rotated by the air-flow across the wing. This use of centrifugal rather than hydraulic forces to produce droplets results in a still fairly wide droplet size spectrum, but without the production of the myriads of tiny drops or the few really large drops so wasteful of insecticide that is characteristic of hydraulic nozzles. In fact, the restricted droplet spectrum of the Micronair means that flow rates as high as 100 l/min are not necessary in practice.

The second development really stemmed from attempts to provide the Third-World peasant farmer with hand-held spraying equipment using very little water. Water, often a scarce commodity during the crop season in the tropics, may also need to be carried a large distance. Contemplating reducing applied volumes to perhaps only 1 l/ha forced a new approach to the production of droplets, since, clearly, large drops had to be eliminated from the spectrum. To break up 1 l to cover a hectare effectively leaves little margin for any wastage! The machine eventually developed, the spinning cup (Fig. 1.6), involved the break-up of rods of liquid rather than sheets. This is achieved by allowing liquid to dribble under gravity on to the inside of a plastic cup spun rapidly by a battery-powered electric motor. The liquid is spun outwards by

Fig. 1.5   The Micronair spinning cage nozzle.

centrifugal force to form rods along fine grooves machined in the face of the cup. As these rods disintegrate into droplets as they leave the cup, a spray with an extremely narrow droplet spectrum can be produced, and the desired droplet size can be varied over a considerable range by changing the speed of rotation of the cup. Output is however limited by the flow rate the cup can take without flooding and forming a sheet of liquid around the edge. Under the right environmental conditions and in skilled hands, the 10 g of active ingredient on plants which seems to be needed with many compounds to give effective pest control can be achieved with as little as 30 g/ha issuing from the spinning cup. With quantities as low as ½ l/ha now practical, oil rather than water can be used as the carrier to increase the weight and decrease the evaporation of small drops. In theory, the spinning cup is a great stride forward; if drops, because of their more uniform size, can be deposited on plants in large proportion to the total volume applied, it must be possible to use less actual pesticide per hectare. Unfortunately, in practice, the reduction in spray

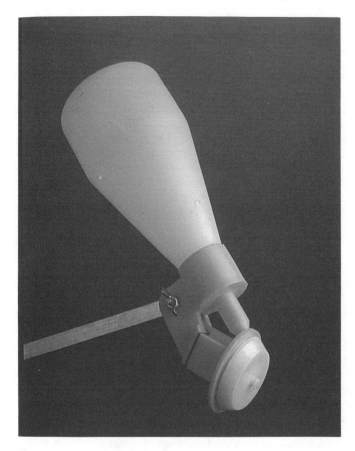

Fig. 1.6   A spinning cup nozzle.

volume obtainable with these devices has not been accompanied by recommen-
dations to reduce the rate of insecticide; indeed, attempts to reduce the rate
have frequently led to failure of pest control. The reason is that with such a
narrow droplet spectrum, if the droplet size is not perfectly matched to the
target and the environmental conditions, droplet capture will be very poor.
Since wind speed, air-crop turbulence, temperature and humidity are just
some of the many factors determining what is the correct drop size to produce
at the rim of the spinning cup, getting it right requires an often impracticable
level of sophisticated measurement of environmental variables and subsequent
interpretation. Perhaps the spinning cup has perfected a technology of droplet
production ahead of our ability to exploit it. The older hydraulic sprayers, with
their wide droplet spectrum, at least have a built-in reliability factor. Under
very different crop and environmental conditions, the 750 g of insecticide
sprayed will always have about 10 g distributed at the right droplet size,
whatever the conditions that day. The Micronair is perhaps as far as we can go

with confidence at present in reducing the spread of the droplet spectrum.

The third development, electrostatic sprayers, really originated with the same motive as that which inspired the development of the spinning cup, i.e. to reduce dependence on water as a carrier needed in large volume. The most successful such sprayer has probably been the 'Electrodyne', into which liquid is fed by gravity between two parts of a nozzle charged at 25 kV. The voltage between the nozzle and an earthed electrode forms an electric field which both atomises the ligaments and charges the droplets. The charged spray cloud induces the opposite charge on the target, and becomes attracted to that target – giving good 'wrap-around' of both upper and lower leaf surfaces. Very small and uniform droplets can be produced; only certain oil formulations can be used, but the low volatility of the oil and the attraction of the spray to the plant result in minimised drift. Volumes as little as ⅓ l/ha may be applied. Penetration of crop canopies is rather poor, however, since the spray is attracted to the part of the plant nearest the spray. However, this characteristic may have a particular advantage in pest management (see Chapter 9). Electrostatic spraying techniques still require further development, and their potential in pest control is not certain.

Additives such as 'wetters' and 'spreaders' are added to the pesticide to improve the 'flattening' and retention of the drop on impact with the leaf. Too much additive will result in the drops coalescing and liquid running off the leaf. The correct amount of additive will vary with individual crops and their leaf surface properties; the *lowest* amount of additive which gives adequate wetting will also give the maximum deposit of pesticide.

Whatever deposit is finally achieved from the inefficiency of the spraying process is immediately subject to erosion with time. Water evaporating from the leaf surface will carry dissolved volatiles away, and falling rain will wash the deposit off the leaf to a greater extent the more that 'wetter' has been used. Wind-borne dust abrades the layer of insecticide; leaf growth and daily expansion and contraction will cause brittle dry deposits to flake off. The deposit is also very vulnerable to chemical degradation; some of this is enzymatic within the plant or by the leaf surface microflora, but especially important is oxidation usually catalysed by the ultraviolet radiation in sunshine ('photochemical oxidation').

Much of what remains of the deposit (now strictly the 'residue') will never be contacted by an insect; even the small proportion of the total residue on the crop picked up by insects may partly be knocked off again as the insects clean themselves or move it to parts of the body (e.g. the wings for many insects) where it will have little effect. This leaves an infinitesimal fraction of the pesticide bought by the farmer on a useful part of the surface of the cuticle of insects. This minute fraction still has to be adequate to sustain the further losses that yet have to occur, and to accumulate in sufficient quantity at the site of action within the insect. The first barrier is the cuticle. Not surprisingly, this external skeleton of the insect is adapted to insulate the insect from the external environment. Insects, as terrestrial creatures, have to protect themselves against water loss; the outer layers of the cuticle, particularly the thin surface wax layer, form an effective barrier to the movement of water. The tissues and body fluids of insects contain many non-aqueous compounds, and the inner layers of the cuticle resist the penetration of such compounds, including oils and waxes. Thus the cuticle can resist penetration by insecticides which are solely soluble in either water or oil. An effective insecticide therefore should have solubility in both oil and water (a good oil:water partition coefficient) and thus be able to move in either phase into different parts of the cuticle. This can be achieved by choosing a suitable solvent.

The site of action of an insecticide is a specific tissue (usually even an enzyme in that tissue) representing a small fraction of the total insect and surrounded by a mass of physiological and biochemical barriers or alternative sinks for the toxic molecule. Many of the non-target tissues (e.g. fat) will store insecticide, and many tissues, including the body fluids, carry enzymes capable of oxidising, hydrolysing or otherwise splitting the toxic molecule. Killing an insect is like trying to fill a bucket full of holes. It is not the amount of insecticide which reaches the target site that is critical; it is its accumulation, i.e. the rate of arrival minus the rate of removal. Thus many times the actual lethal dose must be picked up on the cuticle of the insect for death to follow. The main problems are passage through the cuticle at a sufficient rate, storage in the cuticle and fat, slow diffusion through the skin underlying the cuticle and through the other tissues, lack of penetration of the sheathing of the nerves (often the site of action), which is a barrier to ionised materials, and the breakdown of the toxin by insect enzymes. The last-named is particularly relevant to the insect's resistance to insecticides (see Section 1.3.3).

1.3 Problems of insecticides

The first synthetic broad-spectrum insecticides, the organochlorines, offered a revolution in the efficacy of pest control, especially of the malaria mosquito and other disease vectors of mankind. Additionally they offered cheap, sure and long-lasting control of crop pests. With hindsight, we can be critical of the profligate early use of organochlorine insecticides; however, we need to remember that it is hard to predict the unforeseen. Many of man's technical advances have in the past needed, and will continue to need, modification in the light of experience. Progress will always involve risk. The wide-scale use of the early synthetic broad-spectrum insecticides, particularly certain organochlorines with the then-prized characteristic of long persistence, produced certain obviously damaging side effects. The scientific community, operating via committee structures and accurate but rather conservative memoranda through official channels, was drawing the attention of governments to these problems as early as the mid-1940s. However, the real landmark in change of attitudes and government legislation was the publication of Rachel Carson's *Silent Spring* in America in 1962. This castigation of insecticides was considerably overstated in the opinion of many scientists, but it created a public awareness and outcry of which politicians could not help but take notice. The problems of insecticides are discussed below under a selection of

Rachel Carson's own chapter headings, which in themselves give some impression of the flavour of her book.

1.3.1 Elixirs of death

Here Rachel Carson pointed out that the synthetic insecticides in regular use had very non-specific toxic mechanisms which rendered them poisonous to humans as well as to other animals, including the insects.

The dilution and spraying of poisonous chemicals present an obvious hazard to the farmer and grower. Rough estimates based on World Health Organization data for 1972 suggest perhaps 200 000 cases of such occupational accidental poisoning by pesticides in that year, though only 1% of the cases resulted in death in countries where suitable medical treatment and antidotes are readily available. However, these figures compare well with the industrial risks of many of man's other occupations. Accidents are often the result of ignoring safety procedures, which have been established for handling pesticides just as for other industrial operations. It is sometimes hard to reconcile safety procedures with practice; for example 'fully protective clothing' may be both too expensive and uncomfortable for a peasant farmer in a tropical climate!

Other fatalities arise from ignorance or negligence – such as storing surplus insecticide in used soft drink bottles or using insecticide containers for storing water – and other fatalities represent murder and suicide. In these respects insecticides are just as prone to misuse as other toxic chemicals (aspirin, disinfectant, etc.) regularly stored in the home.

A concern more specific to pesticides is any danger attached to the chronic regular intake of small quantities as residues in our food. Like radioactivity, pesticide levels decrease with time along a hollow curve. If half the pesticide has gone in 5 days, then ¼ will still be there after 10 days, $1/8$ after 15 days, $1/16$ after 20 days, and so on. Great alarm followed the discovery in the 1940s that post-mortems of, for example, accident victims revealed measurable quantities of DDT in the body fat of nearly all inhabitants of developed countries and that no restaurant meal could be found free of DDT residues in such countries. The alarm was heightened by the linear increase of these DDT residues in body fat year after year. However, it could later be shown that man eventually attained a plateau level of about 10 parts per million (ppm) DDT in body fat, and that thereafter further intake was balanced by elimination from the body. More modern pesticides also lead to residues in our food, but are much more quickly broken down and lost from our bodies. Nevertheless, it is still reasonable to ask 'Do such residues do us any harm?' It is obviously impossible to guarantee that they do not, and there has been no shortage of effort to try and demonstrate that they do. The failure of such efforts in spite of the fact that man has been imbibing insecticide residues for over 40 years does suggest the risk is small when compared with the gain in food supplies insecticides have achieved for us. However, alternative ways of protecting our crops or at least reducing the amount of pesticide used are worth serious consideration if they have the potential to reduce the risk still further.

1.3.2 And no birds sing

One of the most publicised side effects of organochlorine pesticides was the death of many birds. The Penguin edition of *Silent Spring* makes this point with its colour picture of a dead bird on its cover. This death of birds had two main causes. First, the compounds were being used widely as seed dressings for grain which was then eaten by birds; second, pesticide applied to crops by drenching the soil leached into waterways, where it was concentrated in particular water layers, and through food chains until the higher predators became affected. Predatory land birds such as eagles, kestrels and hawks as well as water fowl, particularly grebes, all suffered from acute organochlorine poisoning. One of the most dramatic examples of this effect was seen at Clear Lake, California, where organochlorine pesticide (DDD) was added after a very careful calculation of the lake volume was made to ensure that no more than one-fiftieth of a part per million was present. The purpose was to control biting flies which were interfering with the use of the lake for recreation purposes. Large numbers of grebes were killed, and post-mortem analyses showed that they had accumulated DDD to the high level of 1600 ppm.

The human race is peculiarly devoted to its 'feathered friends', and this kind of side effect of pesticide use had no difficulty in reaching the headlines. However, there are also many less-publicised effects on lower animals which represent equally if not more devastating repercussions of the broad-spectrum toxicity of most pesticides.

One of these is the so-called 'resurgence' problem. A broad-spectrum chemical may be highly effective against a highly mobile pest, but at the same time be equally or even more effective against other organisms, particularly insects beneficial to crops. In general, the sensitivity to pesticides of these beneficial insects is considerably higher than that of the pests. The mobile pest reinvading the crop is then able to multiply without restraint from its natural enemies, and a far worse pest problem may result than was present before the pesticide was applied. Already quite a long time ago, Ripper (1956) was able to cite many examples of the resurgence problem, including the classic one when the chemical para-oxon created the 'most enormous cabbage aphid outbreak . . . ever . . . seen in England'. Such spectacular resurgences are often short-term and easily reversible, but there are other examples of slower, less spectacular but usually much more permanent phenomena. The appearance of new pests as the result of the use of pesticides is usually a long-term effect. The classic example of such a man-made pest problem is probably that of the fruit tree red spider mite (*Panonychus ulmi*), which was promoted from insignificance to major pest status in the 1940s. The problem resulted from the use of DDT against codling moth and the use of tar oils against over-wintering eggs on apples. DDT killed the red spider mite's predators as well as stimulating its fecundity. Tar oil killed the eggs of predators and competitors, whereas red spider eggs selectively survived because they carry their breathing pores on an elongated tube rather like a snorkel which penetrated through the oil film.

There are at least two non-target organisms in the crop other than natural enemies. First, bees are often important pollinators and are at considerable risk from insecticides. Although nearly all farmers and growers either avoid spraying when their own crops are in flower or warn bee keepers to close their hives, bees may suffer from insecticide drift, which can carry insecticide on to fodder crops and wild plants on waste land or hedgerows near the crop. Second, there is the effect of pesticides on the crops they are designed to protect. Sometimes the symptoms of this phytotoxicity are spectacular, as with the sudden leaf drop of 'sulphur-shy' currant bushes when a sulphur-containing compound is used. Such dramatic phytotoxicity is usually spotted by the chemical manufacturer during the pesticide's development, and the chemical will then be diverted to the herbicide section. Some other expressions of phytotoxicity are much less obvious, and may only be discovered after the chemical has been in use for many years. Many pesticides cause some check to the growth of the plant and result in small reductions in yield, but farmers are rarely in the position to compare the yield of sprayed versus unsprayed crops. Other examples include an impaired crop flavour (taint), a reduction in fruit set and a tendency to accentuate the biennial cropping problem in apples. Phytotoxicity may also result from formulation. For example, addition of high wetter concentrations to wet the waxy surface of many fruits has sometimes caused local damage where the liquid has run off and left a drop, gradually concentrated by evaporation, at the base of the fruit.

It must be stressed that many of the long-term problems of effects on non-target organisms, particularly the larger 'wildlife', have now been reversed since the highly persistent organochlorines have mostly been replaced by much shorter lived insecticides.

1.3.3 Nature fights back

To some extent greater knowledge and better testing have enabled industry to produce modern pesticides that are safer for humans and less damaging to the environment. The problem with pesticides that has yet defied solution is that pesticides are likely to lose their effectiveness after prolonged use, sometimes even after only a few seasons of use. Pest populations invariably have a genetic pool of widely differing susceptibility to the poison, and the use of pesticides creates a selection pressure on the population whereby the less susceptible individuals are left to breed the next generation (Fig. 1.7). These individuals have properties such as less permeable cuticles, faster storage of toxin in fat or a better equipment of enzyme systems for metabolising the toxin. These properties are genetically inherited and can be passed on to their offspring. Originally rare in the population, the possession of these properties therefore becomes increasingly common, as shown in Fig. 1.7. The appearance of individuals apparently far more resistant to the pesticide than were present

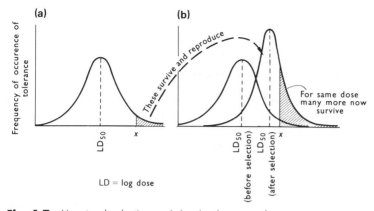

Fig. 1.7 Unnatural selection and the development of resistance to insecticides. (a) Distribution of tolerances before selection. (b) Distribution of tolerances after selection compared with distribution before selection. x, the dose exerting selection pressure, kills fewer organisms in the population of progeny than in the population of parents.

in earlier generations is due to the fact that they are becoming more common and are therefore likely to be included in a sample.

Two questions which perhaps arise in the reader's mind are: 'Do not natural enemies similarly become resistant?' and 'Why cannot we apply a dose of pesticide high enough to leave no resistant survivors to breed the next generation?' The answer to the first question is 'Yes, but much more slowly than pest species'. Predators and parasitoids tend to have fewer generations per year than pests, and therefore fewer opportunities for genetic selection. Also, the pesticide is applied by techniques specifically designed to make effective contact with the pest species. Natural enemies are behaviourally different and may often consume much untreated prey; they are also often mobile and can therefore avoid pesticide better than the pest species can. Moreover, the numbers of natural enemies depend on how abundant the prey is, and therefore the prey has to develop resistance first before a change in abundance of the predator is noticeable. It is therefore not surprising that the first resistant predators found in the field were predatory mites preying on pesticide-resistant mite pests. The behaviour and spatial occurrence of the predatory mites and their prey are also fairly similar. Although the apparent survival of such mites following pesticide application was noticed as early as 1953, it was not until 1970 that the resistance was confirmed in the United States. Experiments have been carried out since 1949 to try to select for insecticide resistance in natural enemies under laboratory conditions, with a view to releasing such resistant strains into crops to be treated with the relevant insecticides. This idea of inducing insecticide resistance in predators and parasites has recently gained considerable momentum. The appearance of pesticide-resistant predatory mites in the field has been followed by further reports of other species. There are at least seven reports of resistant natural enemies in the field, of which five are mites, one a ladybird and one a parasitic fly.

The answer to the second question is that there are in fact rare, remarkably resistant individuals in the population before insecticide is applied. Although an excessive dose would probably kill even them, there is often a relatively small margin of selectivity of action of the pesticide between insect and plant, and there is also a limit to how much pesticide can be brought in contact with an insect.

The 'tolerant pest strain' problem is so serious because there is a real danger that the appearance of such strains will outstrip the production of effective pesticides. As the problem seems to be the inevitable consequence of pesticide use, it is not surprising that all groups of pesticides are affected. Over half the world's pests show a tolerance to at least one major group of insecticides. The production of new chemicals has never been a rapid process; moreover, nearly all the pesticides we use share some similarity, usually in their mode of attacking the biochemistry of the nervous system. The economics of pesticide use mean that there is a ceiling to the price which farmers can pay, and this gives a ceiling to the profitability a new pesticide can achieve before it is forced out of use by the appearance of resistance to it. The costs of testing and developing a new pesticide soar with inflation and increased demand for safety testing. At time of writing, the costs must be approaching £30 million. A new chemical must be protected by patent early in development, and it is not unusual for 6 of the 20 years of patent life to expire before the product is even marketed (Fig. 1.8). Chemicals whose patents have expired can then be

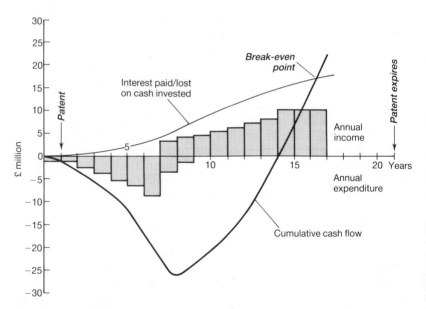

Fig. 1.8   Cash flow in the development of a hypothetical new insecticide, with a break-even point less than four years before expiry of the patent.

manufactured by companies which do not carry the heavy overheads of development, and this will cause a drop in price of the product. The chances of retrieving the development costs with an adequate profit in the short residual patent life and before resistance develops are daily becoming slimmer, and it is in no way surprising that several companies have 'opted out' of the development of new products.

1.3.4 The other road

This chapter comes at the end of Rachel Carson's book, where she urges us to abandon pesticides and seek 'The Other Road', particularly via biological control of pests. The succeeding chapters explore this other road, particularly to see how far a general alternative to chemical control emerges. Rachel Carson pondered what man's fate would be when he came to the end of the insecticide road. By the time her book was published, man had already found the insecticide road to have run out in at least three crop situations. An account of those events, and how they were solved, will be found in Chapter 9.

2

The causes of pest outbreaks

2.1 The pest problem

Estimates of the pest problem on a world scale suggest that, without insect pests, world food production could be increased by about a third. As this estimate represents the loss despite current control measures, it would clearly be catastrophic for mankind if control of insect pests were not attempted or should fail.

Obviously each insect individual has a fairly small food requirement. For example, a greenfly is unlikely to extract more than about 0.5 cm^3 of sap from a plant in its lifespan, and even a voracious caterpillar is likely to consume only 50 g of the fresh weight of its host plant.

Some pests can do damage in relatively small numbers, such as, for example, when one insect blemishes a whole apple fruit or transmits a plant disease. However, the damage done by most pest species results from the enormous numbers in which they occur. There may often be 25 million insects per hectare of soil and 25 000 in flight over a hectare, compared with a human density over the dry land of the earth of about 0.14 per hectare. Obviously, not all these insects are eroding man's food supply; however, numbers of just a single pest species per hectare of crop will often be comparable with such figures. One hectare of oats may harbour 22 million frit fly larvae and 222 million black bean aphids may occur per hectare of sugar beet. Both these infestation figures represent a rapid multiplication of the initial immigrants to annual crops and indeed most insect species which cause pest problems have amazing reproductive powers.

Statistics calculated for such powers would not seem out of place in a science fiction novel. They are impressive, but wildly unreal, ignoring as they do any restraints on population increase of mortality, food supply or a space to live, and in assuming optimum conditions for the whole twelve months of the year! Yet they do serve to illustrate the potential 'population explosion' that pest control must aim to suppress. Thus, the potential of cabbage aphids to produce a new generation every two weeks, with 50 young per female with each generation, is more dramatically (if nonsensically) expressed as a potential in one year of one aphid mother to produce offspring weighing 250 million tons, encircling the Equator nose to tail a million times. Equally startling is the notion that in one year a pair of house flies could cover the earth to a depth of 15 m with their offspring (200 billion individuals).

2.2 Insects outside the crop

Although these days it is hard to find any natural vegetation in Britain of the kind in which insects evolved, inspection of our roadside verges, hedges, commons and woodlands makes it clear that insect herbivores also exist under conditions where they do not occur in vast numbers, and the plants do not appear to suffer extensively from contact with them. We can contrast this 'endemic' situation with the 'epidemic' situation we often find in our crops. It is therefore worth comparing how numbers of insects are affected in contrasting situations, such as uncultivated land and crops, or comparing the same species in countries where it is a pest and in others where it does not attain pest status. The regulation of animal numbers has been discussed by Solomon (1969) and Hassell (1976) in the 'Studies in Biology' series; therefore only a brief discussion will be given here.

2.3 Factors affecting the abundance of insects

The way the numbers of an insect species change obviously represents the balance of births and deaths over a given time period. Birth rate is influenced by many things, including the weather, the food quality received by the adults during development and also the degree of 'crowding' of the individuals. Crowding affects birth rate partly through affecting the quality of the food but also by more direct influences such as stimulating restlessness of individuals. Death rate is influenced mainly by climate and natural enemies or disease; crowding may lead to cannibalism or starvation. Moreover, crowding may also lead to emigration which, like death, leads to a reduction in the number of individuals in an area.

These influences can be classified under two headings: 'Climate' and 'Competition' (Fig. 2.1). However, for the longer term regulation of insect numbers at a particular level, more important than whether a factor is related to climate or competition is whether that factor varies in its impact with density of the insect population it is acting on. This is explained in Fig. 2.2. Because insect populations tend to increase geometrically, a typical population increase can be represented as a straight line plot of the *logarithm* of density against time (Fig. 2.2*a*). If there is a restraint on birth rate, the line will have a shallower slope (Fig. 2.2*b*). If the restraint is very strong, the population may actually decline (Fig. 2.2*c*). Any factor which causes a simple change in the rate of increase of this kind will, if extrapolated, lead either to enormous populations or extinction.

There is clearly a limit to the size of any population if the increase rate is positive (e.g. Fig 2.2*a* or *b*). At some point, density will reach a point where competition for space or food becomes intense and the increase rate will then

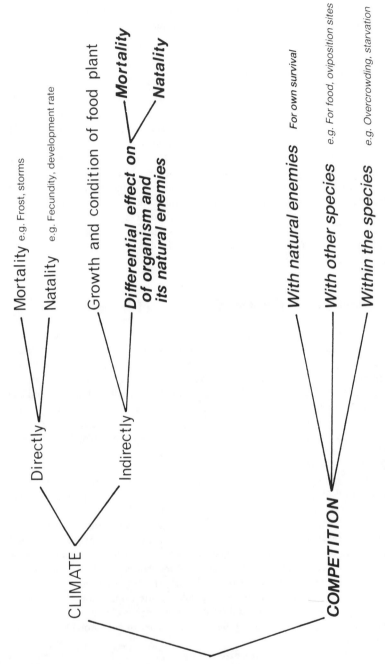

Fig. 2.1 Classification of environmental factors determining changes of insect abundance in the field. Density-independent factors are shown in roman letters, density-dependent factors in italics.

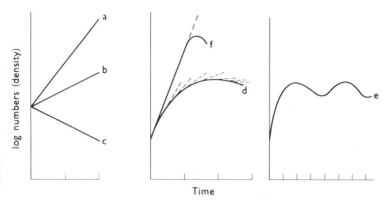

Fig. 2.2 Theoretical plots of the logarithm of insect numbers (log density) against time under the influence of various factors (*a* – *f*) (see text).

slow down quickly (Fig. 2.2*f*). Such slowing down is an example of the action of a *density-dependent* factor; it has little or no impact on the population increase rate until a certain density is reached, when its impact suddenly becomes rapid and dramatic. Until then (Fig. 2.2*a, b* or *c*) the population increase rate has largely been determined by *density-independent* factors. Fig. 2.1 shows the main division into density-dependent and density-independent factors of the various influences of climate and competition.

If our population increase is represented by a straight line of logarithm of density against time, then a density-dependent relationship can be shown by a change of a straight line into a curve. Fig. 2.2*d* shows a density-dependent relationship operating long before overcrowding occurs, and the impact of this restraint becomes greater as the population increases (represented in Fig. 2.2*d* by a series of increasing changes in angle forming a curve). Natural enemies may show such a density-dependent relationship with the density of their prey (Fig. 2.2*e*). The increasing impact of their depredations as pest density rises is partly due to increasing numbers of natural enemies locating the prey source and remaining there to breed and partly due to their spending a higher proportion of their time feeding and less time searching for prey at high prey densities. Furthermore, many natural enemies will spend less time feeding per prey individual when prey is abundant, moving on to new prey without totally exploiting the corpse of their last victim.

Even 'crowding' may exert a density-dependent relationship at a much lower density than would be expected. Encounters between individuals of the same species will lead to some cannibalism at quite low prey densities or stimulate the production of emigrant individuals long before overcrowding is apparent. Clearly natural selection will operate against the dangers to subsequent generations when individuals in an overpopulation situation have to face intense competition for resources. Competition for a convention ('conventional' competition) rather than for an absolute food resource is therefore a widespread phenomenon in animals, including insects. This competition very often takes the form of a territorial behaviour; additionally, increasing contacts between individuals as density increases may reduce fecundity, promote cannibalism and induce emigration or an arrest in reproduction.

Based on the plot of logarithm of numbers against time used earlier, we can represent two contrasting situations diagramatically (Fig. 2.3): situation (a) is where density-independent relationships (Di), such as the quality of the nutrition available, allow the insect to have an increased birth rate fairly close to the optimum (O). Density-dependent relationships (Dd) reduce the increase rate by relatively little as density increases and crowding influences (C) will appear late in time. A second scenario (b) is where density-independent relationships are much more unfavourable, where density-dependent relationships increase their restraint on the increase rate rapidly as density rises and where crowding influences appear early in time.

2.4 Crops versus uncultivated land

In many ways, situation (a) above relates to crops (where insects are often pests) and situation (b) relates to uncultivated land (where insects rarely achieve pest status). The contrast described highlights sources of pest control other than toxic chemicals which are available to man and are discussed in the subsequent chapters, and only hinted at briefly below.

Density-independent relationships: Outside the crop, the susceptibility to attack and the quality of host plants is variable, whereas in the crop, plants have been specially selected and managed so as to emerge, grow and mature in synchrony. Provided that the pest arrives at the right time, it will find a very large proportion of the plants at a high level of suitability. Fertilisers and thinning are among the management practices which maintain a high plant quality suitable also for the rapid development of a pest infestation.

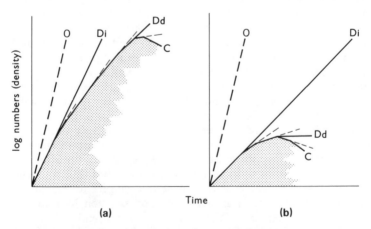

Fig. 2.3   The interplay of density-dependent and density-independent relationships to create high (a) or low (b) insect densities.

Density-dependent relationships (natural enemies): Natural enemies tend to be reduced within crops. There are four main reasons:

(i) Many pests have been introduced from abroad, and have been separated from their natural enemies which may not survive the new climate. The woolly aphid (*Eriosoma lanigerum*), for example, was imported into Britain from the USA; its effective wasp parasite (*Aphelinus mali*) is unable to survive the British winter.

(ii) The continuing presence of vegetation and therefore prey in uncultivated land make the latter a much more stable habitat for natural enemies than is the crop. Most crops are cleared and harvested to leave fallow land, which is then ploughed, resown, and often treated with herbicides to maintain a stand of just a single plant species. Each new crop must therefore largely be colonised by natural enemies from outside, and this may not occur until the prey is in short supply outside the crop or until pest numbers on the crop have built up to an attractive level.

(iii) The use of toxic chemicals, especially insecticides, depletes the natural enemy fauna in crops.

(vi) Natural enemies may have requirements outside the crop for alternate prey or adult food (see Chapters 3 and 7).

Density-dependent relationships (crowding effects): In uncultivated land, individual plants of any one kind are often few and scattered. Arriving insects are met by a bewildering mixture of odour cues which make it much harder for them to locate their host plants, and it is hard for 'overcrowded' populations on one plant to spread to another; large mortalities are likely to occur in the process. In the crop, infestation spreads much more evenly, and average densities can be very high before emigration or other competition effects of crowding cause significant reductions.

The pest population can therefore be crudely compared to a motor car; slow progress can only be maintained uphill (in uncultivated land) by pushing the accelerator to the floorboards (phenomenal reproductive rates). The growing of crops bulldozes the hill away (plentiful food supply of high quality with reduced natural enemy population) – without being able to help themselves, the insects zoom away at a terrifying speed! Many pest insects occur 'naturally' but at rather low population levels, competing for ephemeral niches such as new seedlings appearing following landslips or spontaneous burning. Such species will have evolved particularly high reproductive rates as an adaptation to the losses in transit of the highly mobile adults required for such a lifestyle.

Southwood (1975) has used field data from a large variety of insects to plot the way numbers between generations change in relation to density, and has arranged these plots on a third axis (stability of habitat) to produce a three-dimensional model. This is rather like using the shape of the crusts of mixed-up individual slices of a sliced loaf of bread to recreate the shape of the loaf before slicing! His model combines the contrasts of Fig. 2.3a, b into a single diagram. In many ways, situation (b) of Fig. 2.4 is represented by the endemic ridge, and situation (a) represents the epidemic ridge. Southwood has argued

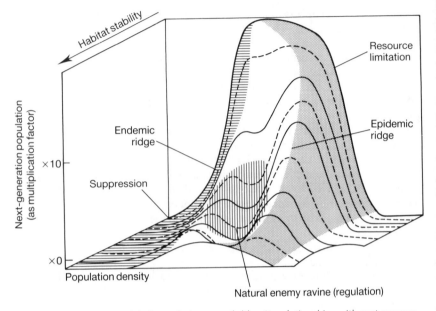

Fig. 2.4   Synoptic model of population growth/density relationships with pest management strategies superimposed. Modified from Southwood (1975), courtesy of Academic Press.

that the really pioneer insects seeking ephemeral niches outside the crop tend always to occur in the epidemic situation with crops and that it is the insects from habitats of intermediate stability which are most affected by the 'natural enemy ravine'. Such insects are normally regulated by their natural enemies in the endemic situation, but any sudden change (especially in density-independent factors such as climate and host plant nutrition) enables density to rise sharply so that the population escapes that regulatory restraint and then will build up irrevocably on the epidemic ridge until eventually the pest population's own responses to crowding slow the population increase at very high densities.

The ideas expressed in this chapter can be exemplified by the documented appearance of a new major pest in comparatively recent history.

In 1824 a British insect collector, Thomas Say, named an attractively striped beetle *Leptinotarsa decemlineata* as new to science. He had returned from an expedition to the Rocky Mountains, where he had found these endemic beetles as scattered individuals on the eastern slopes, feeding on the weed Buffalo-bur (in the potato family). When (30 years later) settlers brought potatoes as a crop to the region, the beetle discovered a new nutritious and well-managed food source on a large scale (i.e. it escaped onto the epidemic ridge). From this time on, the beetle spread eastwards to cause famine at a speed of 140 km a year, and soon reached Europe. Today this collector's rarity, now known as the Colorado beetle, is to be seen all over the world in warning illustrations on police station noticeboards in the company of wanted criminals.

3

Biological control

One of the oldest methods of pest control is the use of other animals as carnivores to reduce pest numbers. More recently, other biological organisms such as insect diseases and plants which are resistant to pest attack have been used for pest control. These are discussed in succeeding chapters, and 'biological control' in this chapter is limited to the use of animals that are natural enemies of insects. For more detail, see Samways (1981) in the 'Studies in Biology' series.

3.1 History of biological control

The use of biological control is probably almost as old as the history of agriculture. Chinese cave paintings clearly show ducks being used to consume pests off crops, a technique still in use in China today. However, the first well-documented case of biological control occurred in 1762, when a Mynah bird was brought from India to Mauritius to consume locusts. In the 1770s, the practice developed in Burma of creating bamboo runways between citrus trees to enable ants to move between the trees more freely for the control of caterpillars.

However, the first real landmark in modern biological control dates from the 1880s, when a ladybird was used to control a scale insect on citrus in California. This example of biological control will be discussed in more detail later. It was this success in California that caused a rush of activity in exploring the possibilities of biological control during this century, and by 1935, 26 successful examples could be listed. DDT, introduced soon afterwards, gave such easy, cheap and successful control that there was considerable disillusionment with biological control which, for most people, was no longer seen as a high priority. However, this euphoria with DDT did not last long, but the result of the DDT era was that by 1958 the 26 examples of 1935 had only grown to 100. The problems created by insecticides soon led to renewed interest in biological control, and by 1964, 225 examples could be quoted. Today there are probably about 400 examples, though it must be pointed out that an 'example' is a local success in biological control, and may well duplicate a system previously employed elsewhere. Nevertheless, probably about 150 species of pest are involved in the successes of biological control reported to date.

3.2 Some examples of successful biological control

a) Cottony cushion scale in California. We tend to think of resistance to insecticides as a fairly recent phenomenon, but by 1887 cottony cushion scale (*Icerya purchasi*) had developed resistance to the insecticides then available, culminating in the failure even of pumping hydrogen cyanide gas under canvas covers put over the bushes. The full story of the biological control programme has many peculiar facets, such as an unfortunate love affair, irregularities with government money and a pair of diamond earrings. The account by Doutt (1958) is well worth reading. By 1988, an American scientist had located two potential biological control agents in Australia. One was a tachinid fly and the other the ladybird *Rodolia cardinalis*. The Americans had considerably more hope of the tachinid than the ladybird, and so the first shipment of insects from Australia comprised 12 000 flies and just over 100 ladybirds, though another 380 of the latter were sent later. However, it was the ladybird (Fig. 3.1) that gave complete success only 15 months after arrival of the first shipment, and it has since been used in many other places against the same pest.

b) Coconut moth in Fiji. This example is cited for several reasons. It is another early example, but also illustrates the successful use of the kind of biocontrol agent which was not successful on citrus in California, a fly in the family Tachinidae; it is cited also because it involved the transfer of the biological control agent to a new host. In the 1920s, British entomologists imported the tachinid fly *Ptychomyia remota*, which normally parasitises another moth, to combat the coconut moth *Levuana iridescens*; 32 750 parasitised larvae were released in Fiji, and control over the whole island was obtained in two years. No doubt success was due partly to the mild climate enabling generations of the coconut moth to overlap continuously so that caterpillars were always available for parasitisation, but also to the fact that the island is fairly small and isolated.

c) Glasshouse whitefly in Britain. This example is included because it is one of the few early British examples. In 1926 a gardener in Hertfordshire noticed that his whitefly scales were not the normal translucent white colour, but black. He sent these black scales to Cheshunt Research Station for identification, where the scales were found to have been parasitised by a small wasp, *Encarsia formosa*. Cheshunt bred and released the parasite to growers from 1929 until the 1940s, when the use of DDT ended the practice. However, the use of *Encarsia formosa* was resurrected in the 1970s as part of a biological control package in glasshouses (see Section 9.2.3). The technique needs good temperature control in the glasshouse, for the parasite is rather sluggish at low temperatures and really too effective at high temperatures when it overtakes the whitefly and then dies out itself.

d) Cassava mealybug in Africa. This is one of the most recent examples of successful biological control. The mealybug was accidentally introduced and

Fig. 3.1 A famous success in biological control: the vedalia beetle (*Rodolia cardinalis*) feeding on cottony cushion scale (*Icerya purchasi*). Photograph from DeBach (1964), courtesy of Chapman and Hall.

spread to 30 countries in Africa. A search in the presumed area of origin of the mealybug, South America, led to the specific parasite *Epidinocarsis lopezi* being introduced to Nigeria in 1981. Since then it has been released at 30 sites, and by 1986 it was established in 13 countries.

Small parasitic wasps, referred to above in connection with the control of whitefly and cassava mealybug, are all members of the Terebrantia (Fig. 3.2) in

Fig. 3.2 The parasitic wasp *Aphelinus* attacking the aphid *Aphis gossyppi*, a pest of cucumbers. Photograph courtesy of the Institute for Horticultural Research, Littlehampton.

the order Hymenoptera (Ants, Bees and Wasps). They are frequently the agents released in biological control programmes.

Other examples of successful biological control are cited later in the discussion of problems that need to be considered when undertaking a biological control programme.

3.3 Advantages and disadvantages of biological control

In biological control, we use natural organisms which have a long history of evolution, much of it before the advent of agriculture. We can therefore assume that the carnivores alive today have succeeded in not over-exploiting their food supply over this long period. The co-evolutionary forces on both prey and natural enemy will have led to neither species being endangered by the carnivorous activity of one of them. Most examples of successful biological control therefore relate to situations where biological control works well enough to keep pests at grower-acceptable levels, where the pest and natural enemy occur together, but where the importation of the pest to a new area without its natural enemy, the use of insecticides or other interferences has led to the herbivorous species becoming a pest problem. Biological control then involves restoring what has been the natural balance, though (see Section 3.4.1) it may often be possible to enhance the effect of the natural enemy in various ways. However, it is also possible (again see Section 3.4.1) that for various reasons biological control cannot be transferred so easily into the new environment.

Many herbivorous insects on crops never become pest problems; the fact that natural biological control may often be responsible has rarely been investigated. However, in many other situations natural biological control is inadequate to keep a herbivore below pest status. It may be necessary to introduce a 'trick' to get the system working unnaturally, although we must then expect the system to evolve counter-adaptations to the trick we have introduced. Importing additional natural enemies may not be as effective as attempting to trick the existing system. Examples of the kinds of tricks which may be employed can be found in Section 3.4, 'The Techniques of Biological Control'.

3.3.1 Advantages of biological control

a) *The technique is selective with no side effects.* Biological control agents tend to be fairly prey-specific, and obviously do not carry the kind of environmental dangers associated with insecticides. This does not mean that side effects can be totally excluded, although they have been very rare in the history of biological control. On one occasion a serious disease of sugar cane was introduced into Trinidad on the ovipositor of a parasitic insect being brought in for biological control purposes, and there have been at least two cases where, after controlling the intended prey, biological control agents have switched to other related herbivores which were important in controlling weeds. Some kinds of side effects on other insects are almost inevitable; the success of the parasite against cassava mealybug (see Section 3.2) has resulted in a decline of the local ladybird predators of the mealybug, because the parasite caused such a dramatic reduction in their food supply.

b) Biological control is cheap. It rarely costs more than about £2 000 000 (as compared with £30 000 000 to develop an insecticide) to complete a biological control exercise, and often this cost only has to be met once. Moreover, it is usually free of charge as far as the farmer is concerned and may be the only economic solution for some forestry and pasture problems, and for many tropical crops grown which have low inputs and are unable to carry the cost of an insecticide. In such situations, biological control does not have to compete with the levels of control given by insecticides.

c) Biological control agents are self-propagating and self-perpetuating. Ideally, once introduced, biological control agents will persist in time, and may spread over large areas from the points of release and reach targets that chemicals cannot (such as larvae concealed in fruit, stems or underground).

d) The development of resistance of pests to biological control is unlikely. Although insects are often capable of defence against attack by carnivores and may, for example, exhibit escape behaviour, release repellent chemicals and encapsulate foreign bodies such as parasite eggs, an existing natural enemy of a pest is clearly already adapted to such behaviour and is, moreover, capable of further adaptation.

3.3.2 Disadvantages of biological control

a) Biological control limits the subsequent use of pesticides. Where biological control agents are being used against one pest, it is clearly difficult to continue using insecticides against other pests on the same crop. This may make the use of biological control impossible unless biological control systems can simultaneously be set up against other pest insects (see Section 9.2.3).

b) Biological control acts slowly. It obviously takes some time for biological control agents to spread from their points of release, to build up in numbers and to make their impact on the pest population. During this period, when the pest may still be present at intolerable levels, any use of pesticide against it or other pests on the crop can endanger the biological control system.

c) Not an exterminant. A biological control system, if intended to be self-perpetuating, involves the presence of the prey even if only at low levels. Growers cannot therefore expect to have a totally clean crop as they can with insecticides, and there may be several types of pests (e.g. blemishers of quality, disease vectors) which, even low levels, will still cause economic damage.

d) Biological control may be unpredictable. The grower has relatively little control over a biological control system, and he often may be worried by this. Even working programmes can suddenly fail. The ladybird (*Chilocorus cacti*) was released against mulberry scales in Puerto Rico in 1938. The control was successful, but over a long period the ladybird virtually exterminated the scale and then died out itself. A sudden mass outbreak of scale recurred in 1953.

Similarly, in the biological control of whitefly in glasshouses, a sudden change in weather or a period of extreme hot or cold can cause a breakdown of the system.

3.4 The techniques of biological control

3.4.1 Inoculation

The natural enemy is liberated in relatively small numbers in the hope that it will establish itself. This approach has been used particularly, though not exclusively, for the control of imported pests and it could equally follow a vacuum in natural enemy activity created after spraying. It has proved especially useful on perennials, against sedentary pests and in 'ecological islands'. For inoculation to be an applicable technique, it is important that the pest should not cause damage at low densities and be known to have arisen as an insecticide side effect or to have been introduced without its natural enemies. Inoculation is also particularly applicable where the problem is widespread and the crop needs little insecticide against other pests.

A survey of possible natural enemies is often carried out in that area of the world assumed to be the centre of evolution of the pest species. It is often hard to identify which natural enemy abroad is likely to be the most effective back at home. There is much merit in seeking a new type of attack; for example, if natural enemies are mainly attacking larvae at home, then a pupal parasite found abroad may well be effective. Also, there might be particular hope for a biological control agent which is closely related to a type which succeeded elsewhere in the world on a similar pest problem. The most abundant natural enemy found abroad may certainly not be the one that is required. The pest species in its centre of origin is likely to be at non-economic levels; a natural enemy required for importation is one which will bring high levels of the pest down to such levels and therefore may appear to be quite rare in the area of the survey.

In general, therefore, it is necessary to ship a large variety of potential natural enemies back to the home country, where they are kept in strict quarantine conditions to eliminate diseases and other sources of mortality as far as possible. Here they will be reared in cages, often on unusual food, and there is usually a decline in the imported population before numbers begin to build up to the point where release is contemplated. Since a relatively small sample will have been imported, and then forced through a genetic bottleneck of culturing, the gene-pool of the released individuals will be very small compared with that of the parent population abroad. This would suggest that the released population is likely to be rather unadaptable, and this may account for several failures of establishment that have occurred following release. However, should the released population establish satisfactorily, the reduced gene-pool may actually be an advantage in limiting the ability of the pest and natural enemy to co-evolve to a higher pest equilibrium level.

During the quarantine phase, a number of studies will be carried out on the selected agents to assess their potential suitability in the field. In general it is desirable that the species should have a high searching capacity, so that they will not emigrate readily when host numbers decline. A host-specific species is likely to be more effective at low pest densities, but will be more prone to ecological interferences such as harvest of the crop. A high reproductive rate is desirable, and is particularly useful if, like many small parasitic wasps, the species has asexual reproduction. It is also important that the species is adapted to the range of climatic variation it is likely to encounter in the field. Here a very good example is the biological control of the walnut aphid (*Chromaphis juglandicola*) in California. This was thought to have originated from Europe, and in 1961 the parasitic wasp *Trioxys pallidus* was introduced from France. This parasitoid established successfully on the coastal plain of California but not in the main walnut growing area of the Central Valley. This is an irrigated desert area, with considerable temperature extremes between summer and winter. A second search for natural enemies therefore centred on the Central Plateau in Iran, which has a temperature cycle similar to that of the Central Valley of California. *Trioxys pallidus* was also found here, and obviously was a different ecotype of the species. The importations from Iran in 1968 were established successfully in the Central Valley.

As well as studying climatic adaptability, it is also important to study the relationship of the development and voracity of the natural enemy to temperature. This will determine whether the biological control agent can cause mortality sufficiently early in the pest's annual cycle to be effective and whether it can then subsequently avoid being 'outstripped' by the pest. It may even be worth looking for natural enemies in a cooler area than the home country in the hope of finding an ecotype with a low development threshold to advance the time of its appearance in the pest population during the season.

Tests also need to be carried out to determine how safe it would be to release the natural enemy in the field, in relation to possible alternative prey, which may already be beneficial as a biocontrol agent of another insect or particularly a biocontrol agent for a weed.

To establish a natural enemy in the field involves trial releases in a number of contrasting habitats; usually 1000 to 5000 individuals are released annually over a 5-year period. The success of the release then has to be monitored over a period of several years to observe whether the pest is declining and whether natural predators are attacking the biological control agent heavily. Here a good example is the largely successful biological control of the Kenya mealybug (*Planococcus kenyae*) in Africa by the wasp *Anagrus* sp. Where coffee is grown by the more modern multiple stem system, a wasp (*Pachyneuron*) attacks *Anagrus*, and this has caused growers to revert to the older single stem system in some areas.

If, after a number of years, only partial biological control is observed, it should not be abandoned, but studies should be undertaken to see whether the biological control agent can be supported by a secondary measure (such as cultural control or the introduction of a plant variety less susceptible to the pest).

A special use of inoculation has been attempted in glasshouses, particularly to control the red spider by the predatory mite *Phytoseulus* (Fig. 3.3). This is based on the concept that the natural distribution of the pest in localised aggregations is disadvantageous to biological control; there is probably no reason why the technique should not be attempted in annual crops. First the pest is inoculated evenly across the crop before natural invasions arrive, and the predator is then similarly inoculated evenly but after a suitable time interval. Even if biological control is not perfect, it may considerably delay the time it takes the pest to reach damaging levels, and the whole system can be wiped out with an insecticide when biological control finally fails.

3.4.2 Inundation

This is the use of biological control as a biological pesticide! Large numbers of the natural enemy are reared in the laboratory and liberated onto the crop; indeed, there are a number of companies culturing and selling biocontrol agents for this purpose in glasshouses. The aim is to create an outrageously high ratio of biological control agents to pests so that the pest is exterminated, the biological control agent itself dies out, and pesticides can then safely be used. It is a technique which has particular appeal where a pest population has

Fig. 3.3 Rearing predators for biological control: dispensing *Phytoseiulus*, a predator of glasshouse red spider mite, from the culture into gelatin capsules for release in glasshouses. Photograph courtesy of the Institute for Horticultural Research, Littlehampton.

become resistant to the available insecticides. However, black scale (*Saissetia oleae*) has been controlled in Zanzibar since 1960 by inundative releases of the imported parasite *Metaphysus helvolus*. The programme involves one to three releases per year.

In California, predators of aphids have been persuaded to inundate lucerne fields. The natural enemies are attracted into the fields by an odour emanating from aphid honeydew, and therefore normally do not invade the fields until large aphid populations have developed. However, the odour from honeydew can be mimicked by spraying the crop with a waste product from brewing, a yeast hydrolysate called 'Wheastrel'.

3.4.3 Conservation

Today there is ever-increasing emphasis on maximising the activity of indigenous natural enemies by either avoiding their large-scale destruction when insecticides are used or by improving the environmental conditions to enhance their survival and activity. The question of careful and even ingenious insecticide use is dealt with in Chapter 9. Concepts for habitat modification to improve the impact of natural enemies are dealt with below.

a) *Microclimate and crop background.* There have been many reports that high humidities favour biological control. In coffee, the pruning system is often designed to maintain high humidity at the right time for parasites of the serious bug pest *Antestia*. Several sources also report that enhanced natural enemy activity in weedy crop plots as opposed to clean ones has a humidity effect. It may be difficult to envisage the practical feasibility of encouraging farmers to leave their plots weedy, but nevertheless some cereal farmers in Britain have recently taken up the concept of leaving an edge area of the crop unsprayed by pesticides, including broad-leaved herbicides. Although this has been done mainly to provide insect food for game birds on farms, it seems likely that the technique will lead to an increase of the natural enemy restraint on insects on the central area of the crop (see Section 7.2.3).

b) *Alternative and alternate prey.* Mention has already been made of the problem that natural enemies may die or emigrate if they reduce their host population to very low or zero levels. Although it is often regarded as an advantage for a biological control agent to be very host specific, so that it does not attack non-target prey, there are other examples where the presence of other insects to act as alternative prey for natural enemies has been necessary to make biological control self-perpetuating. For example, gipsy moth outbreaks tend to occur only in forests without ground vegetation. In forests with ground vegetation, however, the ground flora supports many caterpillars of other species, which are used by parasites of the gipsy moth at times when the gipsy moth is scarce.

Alternate prey is prey that is required by the natural enemy in order to maintain its survival regardless of the abundance of the pest species. A good example here is the grape leafhopper in California. Its egg parasite, *Anagrus epos*, requires a generation of leafhopper eggs for its overwintering generation, but unfortunately the grape leafhopper itself overwinters as an adult. Biological control of grape leafhopper by *Anagrus* has been achieved by planting blackberries near the vineyards. Here the blackberry leafhopper, which overwinters as an egg, acts as a bridging host for the parasite throughout the winter season. In Great Britain, the diamond-back moth (*Plutella maculipennis*) is heavily parasitised by *Angitia* spp. The parasitic wasp emerges from caterpillars on cabbage in the autumn, whereas the caterpillars spin up as a cocoon for the winter and are not suitable as hosts for an overwintering generation of the parasite. It has been known since the 1930s that *Angitia* must bridge the winter in some other caterpillar, but it was not till the 1950s that the late O.W. Richards located *A. fenestralis* overwintering in a caterpillar (*Swammerdramia lutarea*) on hawthorn. *Swammerdramia* was, till then, just another 'economically-neutral' insect.

c) Flowers. Flowers are important sources of food for many adult natural enemies, particularly the parasitic wasps. Their importance is discussed in more detail later; Russian workers have found it advantageous to grow potted flowering plants in cabbage crops at the ratio of 1 pot to 400 cabbages in order to promote biological control. In the absence of flowers, natural enemies moulting to adults in the crop may have to leave the area to find food before their eggs will mature. A series of articles by George Wolcott in the 1940s tells the story of repeated failures of a biological control programme in Puerto Rico due to the absence of suitable flowers for the adult parasitic wasps that were being released. In a later programme in Mauritius, Wolcott introduced the weed *Cordia* into Mauritius before beginning his biological control releases. Interestingly enough, *Cordia* became a major weed of Mauritius and was later controlled in the 1950s by a different entomologist through the introduction of a plant-feeding beetle from Trinidad to Mauritius. Wolcott therefore appears to have been eventually successful in a biological control programme, though at second hand.

4
Microbial pesticides

The spraying of insect diseases onto crops, 'germ warfare against insects', has quite a long history. The first experiments were carried out by the Russians as early as 1886, and a commercial preparation of a fungus was available in Paris in 1891. Like insecticides, pathogens can be stored for a period, marketed in drums, diluted with water and passed through a spraying machine.

As a form of pest control, pathogens have some quite striking advantages. In contrast to insecticides, they tend to be quite target-specific, leave no toxic residues and are unlikely to stimulate resistance in the target organism. In contrast to biological control, many pathogens are compatible with insecticides and can often be used in combination with them. Pathogens therefore would appear to be ideal for control programmes where it is important to maintain the survival of natural enemies. They are clearly also useful for dealing with cases of pest resistance to insecticides, especially where the application of pesticides is restricted, for example close to harvest or on environmental or cost grounds. Thus they have been found extremely useful in some low-input cropping situations such as pasture and forestry.

Unfortunately, pathogens also have very serious disadvantages. Their high specificity means that their development has economic limitations, since specificity also means market restriction. They are living organisms, often with a very short life span in nature, and so it can be extremely difficult to produce them on a factory scale and store them while retaining their virulence. Once applied in the field, they may fail if conditions are too dry or too hot, or if the pH conditions on the leaf surface are outside certain limits. Many of them are also very sensitive to ultraviolet radiation, and so are rapidly destroyed by sunlight. Another problem is that diseases really spread best where the pest population is reasonably high; there are threshold populations, often above those acceptable to the grower, below which the disease will not spread. Although pathogens leave no toxic residues, they do leave the corpses of their victims, and these may adhere firmly to the plant and be very unsightly, forcing the grower into expensive washing procedures before he can sell his produce. Pathogens cause great difficulties in the development process since the direct toxicity which can be demonstrated in the laboratory is perhaps less important than certain behavioural and biological properties of their prey in the field, which ultimately determine the contact between the pest and a sprayed pathogen. Pathogens are of course developed by multiplying diseases first located in natural insect populations, and this may often mean expensive rearing of large numbers of pests to multiply the disease. However, many pathogens are amenable to multiplication by the modern techniques of biotechnology, and these pathogens have received most attention in recent years.

4.1 Contact microbials — fungi

Invasion of insects by fungi results from a spore landing on the cuticle of the insect, and the germ tube then penetrates the cuticle directly. Early work centred on the genus *Beauveria*, especially *B. bassiana*, used particularly against cabbage caterpillars. In general, control with sprays of fungal spores is unreliable, since the spores require moisture to sporulate and really quite high humidities are required for success. An obvious use of fungi is therefore against soil pests, and the fungus *Metarhizium anisopliae* has shown considerable promise for soil application. Fungi have also been used in glasshouses, where high humidities are not so difficult to achieve and maintain. Much research has been done on a fungus, *Verticillium lecanii*, originally found in scale insects. Different isolates of this fungus have proved effective against aphids and white-flies, and formulations suitable for both purposes have been marketed. Fungi are compatible with many herbicides and insecticides, but of course caution is needed where fungicides also have to be applied to the crop.

4.2 Ingested microbials

These pathogens rely on ingestion by their host to initiate an infection, and are therefore adapted to obtaining a fairly resistant stage in the life cycle to enable their survival on a leaf until ingestion by the pest. Such pathogens are therefore rather less humidity dependent than the fungi.

4.2.1 Virus

More than 300 viruses have been isolated from about 250 agriculturally important pest insect species, yet less than 10 are in commercial production. The successful viruses are in the group Baculoviridae, usually either nuclear polyhedrosis or granulosis viruses. A major success with viruses has been the control of the sawfly *Neodiprion sertifer* on forest trees in Canada using a nuclear polyhedrosis virus (Fig. 4.1). There is currently interest in controlling the armyworm *Spodoptera* on wild plants in Africa to keep populations low and prevent its mass migration on to crops following a build up of numbers in the grassland areas where it breeds. In this instance the disease inoculum is being obtained from natural infestations. Viruses can be applied at a very low dosage; for example, in work with naturally diseased caterpillars on cabbages, the disease of less than two infected larvae was applied per hectare.

The development of commercial virus preparations has been somewhat

Fig. 4.1   Larvae of the European pine sawfly (*Neodiprion sertifer*) dying from infection with a baculovirus. Photograph courtesy of Mr C.F. Rivers.

hampered by fears as to the safety of distributing them in the field in case they then mutate to attack humans or domestic animals. However, many virus diseases of insects are quite specific to their hosts and already occur naturally in host populations. Their use would need to be very widespread before the quantities of pathogen sprayed by man outweighed the inoculum available in natural populations. However, it could also be argued that the development of viruses by man for pest control involves the selection of strains of particular potency for particular target organisms and that the release of these in the field is not equivalent merely to an increase of the natural inoculum.

4.2.2 Bacteria

Bacterial preparations available as powders typically contain about a thousand resistant parasporal bodies per milligram. The powders are wetted and sprayed and the pest may then ingest vegetative bodies while feeding. Each vegetative body contains two structures, a spore and a protein crystal. Although the spore often releases some toxins, the important insecticidal element is the protein of the crystal. When the parasporal body reaches the high pH of the gut, the protein crystal dissolves. This is what kills the pest; the spore itself does not germinate and propagate vegetatively until pH conditions change when the gut ruptures and the insect becomes a corpse. As the cadaver dries out, the bacterium resporulates to form parasporal bodies.

Today there is considerable interest in developing the toxic protein carried by parasporal bodies as an insecticide in its own right.

The primary bacterium commercially available is *Bacillus thuringiensis*, often simply known as 'BT'. A fresh commercial preparation of *B. thuringiensis* has often compared favourably with normal insecticides for the control of insects. BT, like most bacterial preparations, requires high humidity for good control. It has the great advantage of being harmless to honey bees and suitable for application close to harvest of edible foodstuffs. Increasingly it is becoming apparent that different strains of BT can be developed to be highly target-specific to different pest species. The strain var. *israelensis* is very toxic and specific to mosquito larvae. By 1980 nearly 30 'varieties' of BT were known, and the primary cause of the variable spectrum of activity of the varieties seems to lie in the spectrum of activity of the protein crystal toxin.

A 'double microbial' has been tried in China to obtain improved control of codling moth. Eelworms were infected with a bacterium and then sprayed. Infected nematodes eaten by the codling caterpillars pierce the gut and speed up the onset of bacterium multiplication in the dying grub.

Another bacterium, *Bacillus popilliae*, is available specifically against the Japanese beetle, *Popillia japonica*, and is mixed into the soil to combat this underground pest of grass in the USA.

4.2.3 Protozoa

Protozoa, especially those in the genus *Nosema*, have also been investigated as insect pathogens for biological control. *N. locustae* has been used successfully against grasshoppers in rangeland in the western USA. A real problem with protozoa is that they can only be propagated on living insect hosts, which makes their commercial multiplication extremely expensive.

4.2.4 Nematodes

Although not strictly an insect 'pathogen', nematodes are drought resistant and small enough in size to be amenable to storage and spray application in much the same way as pathogens. Comparatively little use has been made of nematodes in this way, although they have been applied against the Colorado beetle in Canada; at the moment there is a great deal of research on one particular genus, *Neoaplectana*. *N. carpocapsae* has been shown experimentally to be effective against codling moth, an apple pest. Nematodes seem to be the only possible natural enemy of several important pests (e.g. mushroom fly and thrips).

5

Pheromones

It is becoming increasingly apparent that many behavioural activities of insects (e.g. movements, mating, aggregation, alarm signalling) are under population control via chemical messengers produced by individals in the population and liberated into the environment either as volatiles or in faeces, regurgitated food, etc. These chemicals have been given the general blanket term of 'pheromones' (see Birch and Haynes, 1982, in the 'Studies in Biology' series) and are available to man for manipulating insect behaviour either by the use of caged insects, extracts from insects or synthetic production (the actual pheromone or a chemical 'mimic' thereof). Pheromones have particular advantages for pest control because they are usually highly species-specific, leave no undesirable residues in the environment and are effective in very minute quantities.

The pheromones most used in pest control are the sex attractant pheromones usually produced by the female of the species. These attractants were first identified in moths. Naturalists have always known that male moths were attracted to females over long distances, and scent was always suspected. However, the quantities released are so minute that the confirmation and identification of these scents had to await the development of gas/liquid chromatography. Sex pheromones produced by female insects are now known for many insects in many different families, a recent discovery being the sex pheromone of the aphid.

Most sex pheromones are very specific alcoholic esters. The specificity arises not just from the actual compounds, but also from their isometric configuration, the rate of release and the ratios of the several components often involved in the mixture forming the sex pheromones. Those components which elicit some response on their own are known as primary components, but sometimes at least one (secondary) component has little effect on its own but its addition greatly enhances the attraction.

5.1 Use of pheromones for monitoring pest populations

Increasingly pheromones are being used to determine when pests enter crops, when numbers have built up sufficiently to warrant control measures being taken or to predict the correct timing for such measures.

Fig. 5.1 Simple pheromone trap (based on a plastic paraffin funnel) for monitoring the moth *Heliothis* in India.

Not only can the use of pheromones greatly improve the correct timing of insecticide sprays, but the simplicity of visiting and counting insects in one or two traps in or around crops makes it a tool which is easy for the farmer to use. The traps are usually fairly simple in design, often little more than a polythene bag attached to a paraffin funnel (Fig. 5.1) or a small metal or cardboard roof with a sticky floor on which the insects arriving at the trap are caught. The pheromone may be released by confining a female in the trap, but for most insects where pheromone monitoring has reached commercial practice, a synthetic pheromone is available and is released slowly from a small rubber or polythene capsule in the roof of the trap.

Unfortunately the use of female-produced pheromones means that it is the males that are caught for monitoring purposes, and it is sometimes hard to relate trap catches of males to the population density of females on the crop and even harder to relate them to the subsequent laying of eggs and hatching of larvae. Even for males, the relationship between population density and catch is certainly not linear, since a number of females producing natural pheromone will compete with a trap and so the proportion of males caught tends to decrease as the total population increases. 'Are there any insects?' is always easier to answer than 'How many?' Once a threshold of males caught in the trap has been reached, it may still be necessary to use temperature records and weather forecasts to predict the correct timing of the spray, which may often be directed at larvae hatching from the eggs.

As stated earlier, pheromones are used to monitor many pest species.

Perhaps two examples of monitoring Lepidoptera pests in the UK will suffice. A recent successful development has been with the pea moth, particularly in relation to dry peas in East Anglia. The growers set up two pheromone traps at right angles at one corner of the field and check their traps twice a week. Ten moths in either of the two traps on two consecutive occasions represents the threshold for action, and growers then contact the advisory service for a forecast of the likely day on which spraying should be carried out. More detail of this forecast is given in Section 9.6.2. Pheromone traps are also used in apple orchards to determine the best date for spraying against codling moth. Here a single trap is set up per hectare of fruit, and five moths per trap in any given week constitute the threshold for spraying.

5.2　Use of pheromones for trapping out pest populations

This technique aims to use pheromone traps, often treated with insecticide, to kill enough males at the beginning of the season to reduce greatly the fertilisation of the females. Unfortunately, mathematical models suggest that 90% of the males have to be killed before the next generation would be reduced, let alone brought down to grower-acceptable levels. Thus, for example, attempts to trap out grape berry moth in the USA were not successful. Traps were set out on a 14-m grid, and although the percentage of infested grapes fell from 15.5 to 6.4% as a result of the traps, 6.4% infestation still exceeded what growers could tolerate. Rather more successful have been attempts to use trapping out to control pink bollworm on cotton in the USA, where the economic threshold for the pest is reasonably high. Traps were set out at 12 per hectare in the spring and increased to 50 per hectare later in the season. Attack of the bolls was also monitored, and sprays were applied when more than 10% were attacked. The technique certainly reduced the need for pesticide, which was applied to 45% of fields with no traps, but only to 9% of those with traps.

A sex pheromone is known to be produced by the male boll weevil, another important cotton pest. Here the ability to trap out females makes the technique very successful. Perhaps future research will show that pheromones are produced by the males of other pest species also; after all, it is not so long ago that we thought that only the Lepidoptera produced sex pheromones.

5.3　The pheromone confusion technique

The aim of the confusion technique is to lay artificial pheromone trails or even to saturate the crop environment with the odour of synthetic pheromone, in order to confuse the males and prevent them locating females. For this technique, pheromone traps are increasingly being replaced by smaller pheromone sources such as hollow polymer fibres of only a few centimetres in length or even encapsulated droplets of synthetic pheromone spread over the crop. The latter tend to be limited in value since they only release pheromone for a few days.

In experiments to control plum moth in Switzerland, hollow fibres 20 cm in length were fixed onto stakes in groups of ten. These experiments showed that, although initially the technique failed to reduce the percentage of fruit damage to an acceptable level, damage levels the following year were much reduced since, with the absence of pesticide use, the natural predators were allowed to survive.

Experiments between 1976 and 1978 in California, evaluating the technique against pink bollworm on cotton, employed short 1.75-cm chopped fibres applied at 7 g/ha every 14 days. These experiments clearly revealed one problem with the confusion technique, that control tends to break down at high populations when many natural sources of pheromone exist. None the less, the technique succeeded in reducing the number of fields which needed pesticide treatment, and, even where pesticides were needed, there was still a valuable delay before the first application was necessary.

It seems likely that the confusion technique cannot be relied upon as a single control measure, but it is likely to reduce significantly the number of pesticide applications needed. At present there is little demand for pheromones, so the high cost would probably make it uneconomic to use pheromones if pesticides also need to be used; the picture could, however, change dramatically if more farmers adopt the technique and the demand therefore increases.

6
Plant resistance

6.1 Introduction

The growing of crop varieties which are less vulnerable to attack than others (Fig. 6.1) or which yield well in spite of attack has many advantages. Once such varieties are available, pest control requires no extra labour and is therefore economical; moreover, the environment does not suffer from side effects of the control measure. Until recently, resistant plant varieties were sought energetically only when other means failed to control pests economically. They are thus widely available against plant diseases and plant-parasitic eelworms. Although some resistant varieties have been used against insect pests for many years, recent pressures for reducing pesticide applications have given the subject a new emphasis in pest control.

Plant resistance poses a problem of semantics. The ability of some plants to yield better than others under the pest pressure in a region may not always be due to something clearly identifiable as resistance in a strict sense. As will be discussed later, so-called 'resistant varieties' may be tolerant or even hypersensitive to pest attack; this has led some people to propose the term 'varietal control'. However, this ignores the fact that it is often possible to obtain resistance by other ways than choice of variety; as the term 'resistance' is no less ambiguous than 'varietal control' and is much better known, it is used here.

6.2 Sources of variation

Most commercially used plant resistance is the result of crop improvement by plant breeders. There are now collections (usually as seeds) of the genetic variation of many of the world's major crops. These so-called 'germplasm banks' have assembled material from a variety of sources. These include current commercial varieties and locally grown peasant varieties collected from a wide geographic range throughout the world. Additionally, new hybridisations by the plant breeders are added to the collections, as well as the progeny of random out-crossing trials (where a male sterile variety is allowed to cross at random with a whole range of varieties surrounding it). New genetic combinations may also be obtainable by irradiation or chemical treatment of seeds.

Fig. 6.1 Junction of experimental plants testing for varietal resistance of lettuce to lettuce root aphid (*Pemphigus bursarius*). The impact of variety as a control measure is seen clearly in the contrast between the aphid-susceptible variety 'Mildura' in the foreground and the adjacent resistant variety 'Avoncrisp'. Photograph courtesy of Dr. J.A. Dunn.

Very often, however, the plant breeder seeks genes for improved yield under exposure to pests in the wild parents and relatives of crop plants. At first sight it may seem surprising that man has not selected for resistance in his crop plants by retaining the most productive types in the course of centuries. However, genetics is a science of this century only, and therefore most selection occurred before it was known that resistance may exist in unproductive types and that crossing can often combine resistance with productivity. Moreover, much early selection was for spot characteristics such as large fruit or large grains rather than productivity. Finally, any resistance which was selected early on may have become nullified in the course of time by adapted races in the pest population or by recent importation of new pest species and races.

6.3 The nature of plant resistance

Plant resistance usually involves a quantitative enhancement of a characteristic already present in a plant. Southwood (1973) has concluded that plant feeding *per se* presents an insect with considerable difficulties; he also pointed out that plants will evolve to counteract damaging insect attack. That resistance to pests is indeed a general phenomenon can be seen in the restricted host distribution of most herbivores, and in the fact that small airborne insects, which have little choice where they land, apparently reject and take off again from plants we regard as their normal hosts with frequencies ranging from 50 to 95%. One of the earliest workers in plant resistance to insects (Painter, 1951) developed a classification of plant resistance phenomena which is still extremely useful and widely used. *Antixenosis* (equals Painter's 'non-preference') refers to plant properties which cause avoidance or reduced colonisation by pests seeking food or oviposition sites. *Antibiosis* directly or indirectly affects the pest in terms of survival, growth, development rate, fecundity, etc. The final category, *tolerance*, refers to a reduced plant response (usually in terms of yield loss) to a given pest burden. It is important to realise, however, that resistance is very often a combination of two or even all three of these phenomena. That plant resistance involves new or resurrected gene combinations should not obscure the principle that the insects respond to phenomena more 'tangible' than the internal arrangement of plant nuclei. A variety is only effective through some chemical, physical or anatomical property of the gene combination. It is therefore possible to identify mechanisms of plant resistance and relate them to types of control (i.e. the field expression of the mechanism on the pest population). This has been done in Fig. 6.2, where an attempt has been made to arrange the mechanisms in order of stages in the pest infestation at which they appear most effective. The aim of plant resistance is to reduce the losses in yield caused by pests. This is clearly achieved when no pest attack occurs, when the variety succeeds in *escaping* the attack. More usually, a variety is attacked but may suffer less attack than the susceptible variety, because it is in some way truly *resistant*. Improved yield may sometimes also be gained from varieties showing a high degree of susceptibility. Some *tolerant* varieties seem able to yield well in spite of pest infestation and yet others are so susceptible as to be *hypersensitive* and collapse locally or entirely under attack; this collapse prevents the pest multiplying and spreading through the crop.

6.4 Mechanisms of resistance

For many resistant varieties, the mechanism is unknown or poorly understood, but some mechanisms which have been identified, with selected examples, are listed below.

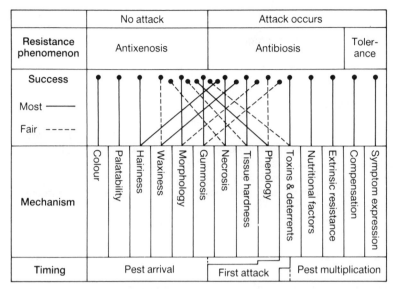

Fig. 6.2 A classification of plant resistance mechanisms based on host plant influences on insect populations. From van Emden (1987), courtesy of Academic Press.

6.4.1 Colour

Insects are often affected by the wavelengths reflected from plant surfaces. The light-green colour of 'Spanish White' onions seems to deter thrips from colonising the plants, and the apple sawfly seems to be attracted by a high ultraviolet reflection from apple flowers and therefore oviposits more on white-flowered varieties. Red cabbage varieties are avoided both by the cabbage aphid (*Brevicoryne brassicae*) and the cabbage white butterfly (*Pieris rapae*), yet the red varieties appear to be more favourable for the reproduction of aphids and survival of the caterpillars.

6.4.2 Palatability at the host selection stage

Very often insects select plants which are botanically related and discriminate for characteristic secondary compounds. It can be shown very simply that cabbage caterpillars will not feed on leaves such as bean or lettuce, but will do so if the leaf surface is painted with mustard oils (rather dilute table mustard solution will do for this purpose), the characteristic flavour compound of plants in the cabbage family. There is no evidence that such secondary substances play any role in the nutrition of the insect; indeed, the same substances deter feeding by many other insects, presumably the vast majority of any particular plant species. Breeding for high levels of these compounds may deter the host-specific insects; breeding for low levels would merely encourage attack by other insects. Introducing a novel secondary compound into a variety normally involves crossings between quite unrelated plant species; this is unlikely to be successful, but modern techniques of gene-splicing may make it possible in the future.

6.4.3 Hairiness

An increased density of hairs on leaves may deter oviposition by small insects. Hooked hairs on certain bean varieties have been shown to trap landing aphids and leafhoppers, but currently most interest centres on the glandular hairs found in wild relatives of potatoes. When small insects such as aphids break the hairs in walking over the leaf, the broken hair exudes a fluid which hardens on the insect's legs and mouthparts.

6.4.4 Waxiness

Several workers have found differences in susceptibility to a variety of pests in glossy and waxy leaf types. Sometimes, as in the case of flea beetles, it is the waxy type which is resistant; for other pests, such as aphids and whitefly, it is the waxy variety which is susceptible. In the latter case, the resistance of non-waxy cereal varieties to aphids has been linked with higher levels of certain chemicals in the plant's cuticular waxes.

6.4.5 Major morphological characters

Here the best examples come from the extensive cotton breeding programmes, largely with the moth *Heliothis* as the target. Some resistance has been obtained by breeding for narrow twisted bracts and the absence of leaf nectaries. Both characteristics reduce the attractiveness of cotton to the moth. Another example is the resistance of cowpea varieties with long peduncles and erect pods to a pod-boring moth (*Maruca testulalis*). The resistance arises because larvae of the borer penetrate the pods most successfully wherever pods are in contact with each other or the foliage (Fig. 6.3).

6.4.6 Gummosis

Many plants protect themselves against wounding by exuding gums, latexes and resins. In conifers, differences in resin flow have been implicated in the resistance of some pine species to attack by pine shoot moth, and some legume varieties produce gum from the pods when damaged; this seems to drown young bruchid beetle larvae attempting to penetrate.

Fig. 6.3 Plant resistance in cowpea to the pod borer *Maruca*. *Left*, susceptible variety with pods touching and drooping; *right*, resistant variety with separated and erect pods.

6.4.7 Necrosis

Here the plant is so sensitive to attack that immediate death of the affected tissues or of the entire plant ensues. A rapid local necrosis wherever aphids probe with their stylets is a known resistance mechanism to some apple aphid species, and even whole plant necrosis can be a resistance mechanism in crops with high plant populations such as cereals, thereby limiting the spread of infestation through the stand. Crops such as cereals show a yield plateau over a range of sowing densities around the commercial rate, and the loss of a proportion of plants from a hypersensitive reaction to attack would therefore not reduce yield per unit area. Recently, considerable interest has focussed on the existence of damage-induced changes in plants which confer resistance to a subsequent pest attack. Chemically these induced changes seem similar to local necrosis, but it is premature to speculate whether this phenomenon can be exploited in resistance breeding.

6.4.8 Tissue hardness

Rapid cutinisation of epidermal cells and rapid cork formation in seedlings often protect fast-maturing varieties from the many pests which concentrate their attack on young tissues. Differences in hardness of plant parts between varieties has been found important in control of cabbage root fly (*Erioischia brassicae*), wheat stem sawfly (*Cephus cinctus*) and the rice stem borer (*Chilo suppressalis*). Work on *Chilo* revealed that it is the silica content of the tissues which blunts the mandibles of the tiny invading larvae, and silica content and distribution is often an important component of 'hardness'.

6.4.9 Phenological resistance

Some varieties appear to be resistant because they are at a less susceptible stage when pest attack occurs, either because of their rate of development or farmer management. Thus early flowering pea varieties escape injury from pea moth (*Cydia nigricana*), although if sown later so that flowering coincides with the pest's presence, they are just as susceptible as other varieties.

6.4.10 Toxins and feeding deterrents

The production by some plants of what may loosely be referred to as 'toxins' has been closely examined in the development of plant resistance to plant pathogenic fungi; phenols seem to be particularly widespread and useful fungistatic compounds in plants. They have also been implicated in the resistance of some apple varieties to root lesion nematode (*Pratylenchus penetrans*), and are probably involved in the local necrosis in response to wounding by insects.

Little use has been made of the toxins produced by plants in antibiosis, presumably because the pests are mobile and show their reaction at the palatability stage (Section 6.4.2). However, survival of larvae of the corn borer *Ostrinia nubilalis* is reduced in maize varieties high in the quinone Dimboa, and grafting experiments have demonstrated the presence of leaf-synthesised toxins in some legume varieties that repel bruchid beetles and aphids.

6.4.11 Nutritional factors

Southwood (1973) regarded the high-carbohydrate and low-nitrogen status of plants as a major hurdle in insect nutrition. In spite of a great deal of knowledge of insect nutrition and how this can be varied through the host plant, little is known of its importance in influencing the resistance of crop varieties to pests. Some information is, however, available on aphids, which mainly feed in the phloem. Phloem contents can be altered without necessarily affecting the value of the plant for man and his domestic animals. Extensive work in Canada has linked the susceptibility of pea varieties to the pea aphid (*Acyrthosiphon pisum*) with the soluble nitrogen content of the plant, particularly the amino acids. 'Perfection' gave much denser amino-acid chromatograms then 'Champion' and on another resistant variety ('Onward') the aphids were comparable in growth with aphids fed on a susceptible variety but starved for 10 hours daily. Work at Reading University has linked resistance of brassicas versus two aphids, the cabbage aphid (*Brevicoryne brassicae*) and the peach potato aphid (*Myzus persicae*), to changes in the amino-acid spectrum. Different amino acids affected the two species and, whilst some amino acids were favourable to aphid increase, others were unfavourable.

6.4.12 Extrinsic resistance

This is the interaction of plant variety with other souces of pest mortality, the most common one being biological control. Probably the best known example is that of open-leaved crucifer varieties, which make it much easier for parasites to find their cabbage caterpillar hosts. An interesting example has recently come from Australia, where the ladybird *Cryptolaemus*, which normally feeds at cotton leaf nectaries, becomes a predator of *Heliothis* caterpillars on the nectary-less varieties.

6.4.13 Compensation

When whole plants or different parts of the same plant compete with one another, the plants or plant parts surviving attack can grow larger. Thus early injury to cereal crops by wheat bulb fly is compensated for by growth in the plants around the victim of fly attack. Full compensation for leaf damage can occur when leaf area exceeds the optimum required for providing assimilates for the marketable unit (e.g. root crops). When the reverse is true, that is, that the marketable unit is produced to excess in relation to the source of assimilates (e.g. many fruit crops), natural shedding is likely to occur. Some replacement of this shedding by insect attack will therefore not necessarily affect the final yield.

6.4.14 Symptom expression

One of the classic examples of tolerance to insect attack is the reaction of some tea varieties to the shot-hole borer (*Xyleborus fornicatus*). Here the damaging symptom is that branches attacked by the borer break off as tea pluckers pass between the bushes. The tolerant varieties prevent this symptom by producing a wound-healing support for the affected branch. Another example relates to some wheat varieties, which do not respond to aphid attack by twisting and curling their leaves, the damage that occurs in most varieties.

6.5 The evaluation of plant resistance

This normally begins with exposing a collection of varieties to pest attack, either in the field or glasshouse, and monitoring the levels of damage, often simply by visual assessment. A susceptible variety is usually planted at intervals in the trial to allow easy comparison between the varieties to be tested and a known level of susceptibility. When large numbers of varieties are to be tested, as are often available from germplasm banks, seeds will usually be in short supply and the aim of the first trials will be merely to reject varieties which are clearly susceptible. Eventually, as the number of varieties with possible resistance is reduced, larger plot trials are undertaken, and the yields of each variety, treated and untreated with insecticide, are compared. Clearly, the more similar these yields, the more resistance the variety is likely to possess. Eventually the most promising varieties are retested at a number of stations, covering as far as possible the various areas in which the crop is principally grown, while more detailed studies are made to discover more about the nature of the resistance. It is important to identify how far antixenosis, antibiosis and tolerance have contributed to the field effect. These studies are often conducted in the glasshouse. It is important that antixenosis should be strong enough to prevent colonisation by the pest in a 'no-choice' situation. Tolerance can be a dangerous form of plant resistance when pests are fast breeding, because farmers growing such varieties would no longer suppress pest numbers and would thus create an uncontrolled reservoir to infest other varieties. The most desirable form of plant resistance is almost certainly antibiosis, especially if accompanied by some antixenosis and even tolerance. Whenever possible, antixenosis and antibiosis of a variety should be retested with pests cultured for some time on that variety and also with different strains of the pest collected from other areas.

6.6 The transfer of genetic resistance

Resistance is often discovered in an agronomically unadapted variety, and a plant breeding programme is then employed to attempt to transfer the resistance into agronomically more desirable and higher yielding varieties. Progeny from the plant breeding programme will require continual retesting to monitor whether, and to what extent, the resistance has been retained. This is a long process, and few resistant varieties are released until 15 years after the resistance was first discovered.

In perennials, resistance may frequently be conferred by grafting a susceptible scion on to a resistant rootstock. The classic example here is the control of *Phylloxera* (an aphid-like insect) on the valuable grape varieties of Europe by grafting them on to resistant American rootstocks.

Usually, however, a programme of hand-crossing between the flowers of varieties, followed by the harvesting and growing of seeds, is required. The details of such crossing programmes are outside the scope of this book; often pest resistance is only one of several plant attributes requiring incorporation into good agromonic varieties in the same breeding programme. Such programmes usually involve forming hybrids between the resistant variety and the variety with the desirable agronomical characteristics, and then back-crossing these hybrids to the latter parent.

6.7 The problems of using plant resistance

6.7.1 Yield penalty

Most mechanisms of plant resistance appear to involve some diversion of resources by the plant to extra structures or production of chemicals. Thus it is by no means certain that any gene for resistance can be incorporated into high-yielding varieties without some sacrifice in yield. Most entomologists working in plant breeding programmes will confirm that they find resistance in varieties with relatively low yields, and that later the high-yielding progeny of breeding programmes they are asked to retest all appear highly susceptible! There is, however, little quantitative evidence that plant resistance cannot be achieved without a yield penalty, but the matter is sufficiently important to warrant further research.

6.7.2 Problem trading

It commonly happens that resistance to organism A is linked with susceptibility to organism B. Thus one student practical class in the Horticulture Department at Reading University ranked apple varieties in order of resistance to red spider mite, and exactly the reciprocal order was obtained by another class examining the distribution of apple mildew disease! Any anatomical or physiological change in a plant is likely to have a different effect on different pests or diseases. Several examples are known from the pest scenario; smooth-leaved soya beans resistant to the moth *Heliothis* are more damaged by leaf-hoppers, whereas cotton with narrow twisted bracts resistant to the same moth is highly susceptible to plant bugs.

Pyramiding resistance to combine resistances to different organisms may be possible provided that the mechanisms are not incompatible (e.g. hairy and smooth foliage).

6.7.3 Biotypes

In theory, resistance of a plant variety is no more a permanent control of an insect pest than is an individual pesticide. Both control measures exert a selection pressure on the pest, and the minority strain not affected by the control may then become more common. 'Breakdown' of resistance, by the emergence of tolerant 'biotypes', is familiar to plant pathologists (e.g. cereal rusts) and nematologists, and has often appeared as rapidly as resistance to chemicals. However, the biotype problem has arisen remarkably infrequently with insect pests. There are about 20 examples which could be cited of the biotype problem in relation to plant resistance to insects; however, most examples derive from purposefully testing other strains of pest rather than any breakdown of the resistant variety in the field. Reasons for the relative rarity of the biotype problem with insects as compared with that of pathogens probably include the frequent association of antixenosis with antibiosis, the presence of several mechanisms operating simultaneously even where one mechanism was originally proposed and a relatively slow insect-generation time compared with pathogens, decreasing the frequency of selection. It is worth noting that the biotype problem has arisen most frequently with aphids, among the fastest breeding pests.

6.7.4 Variability of resistance

Plant resistance is the result of an interaction of insect behaviour and physiology with definite plant characteristics. In as much as these resistance characteristics are variable according to the age of the plant and the environment in which it is grown, a 'resistant' plant may show susceptibility under certain conditions.

Plants often show a peak in susceptibility to pests at a fairly early age, with young tissues and young seedlings less vulnerable and a steady increase in resistance as the plants age after the susceptibility peak. Aphids then show another very rapid rate of increase as soon as plants begin to flower. Thus it is important to identify the appropriate stage for testing any plant material for resistance.

Many environmental factors have been reported as modifying plant resistance phenomena. Resistance is sometimes lost at either low or high temperatures, or at low light intensities. Major plant nutrients, pesticides and other chemicals such as growth regulators can also affect plant resistance. All this has bearing on the reliability of glasshouse studies of plant resistance, and the importance of confirming resistance in a range of environments. However, the positive aspect is that there is enormous scope for obtaining resistance by techniques other than plant breeding. Obvious possibilities for inducing resistance are fertilisers and plant growth regulators. Such induced resistance circumvents all the problems and delays of incorporating resistance with other desirable plant characteristics in the plant breeding programme.

6.7.5. Plant resistance and vectors of plant diseases

The fact that aphids and other vectors of plant diseases may be more restless and show increased probing on resistant varieties, has frequently led to suggestions that plant resistance could spread, rather than reduce, virus transmission (especially of the non-persistent type). However, no field example can be cited where the introduction of a variety resistant to a vector has increased the spread of a plant disease, although it has been possible to show this effect in cage experiments. By contrast, there are several examples from the field where disease resistance has clearly been reduced even on varieties only partially resistant to the vector.

7

Cultural and legislative controls

7.1 Cultural control

Before the advent of the modern synthetic insecticides, man's chief weapons against insects were to try to disrupt their life cycles by periodically denying them their food plant and to achieve the maximum control that the manipulation of ordinary agricultural practices would allow. Such 'cultural' measures gave control inferior to that given by modern insecticides; they were also labour-intensive and therefore became highly expensive as labour costs rose. Moreover, many were obstacles to farmers' aspirations to intensify and further mechanise their holdings. As a consequence, the availability of cheap and efficient synthetic insecticides caused farmers largely to abandon cultural controls, particularly in temperate countries where labour was becoming so expensive. However, there are now attempts to resurrect some cultural controls as part of pest management programmes (Chapter 9). In the tropics, where peasant farmers often cannot afford insecticides and labour is still cheap, cultural control is still a major pest control weapon. If the future mechanisation and intensification in the tropics follow the European and American pattern by which cultural control measures are ignored, it is unlikely that pesticides alone can be used to cope in the way that they are in temperate areas. The high potential pest pressure of insects, often breeding rapidly year-round in warmer temperatures, is likely to lead to tremendous cost and an unheard-of frequency of application of insecticides, leading inevitably to resistance problems. Cultural control, though providing control inferior to that of pesticides, is a valuable restraint on the average pest density, and therefore is valuable in reducing the challenge that insecticides may be called upon to meet in the future.

Increasing the 'resistance' to pests of the agro-ecosystem has problems in some ways comparable to those of using resistant crop varieties. There may be a yield penalty to consider, as well as the potential for 'problem trading'. Cultural control aimed against one pest may well improve conditions for another. The potential impact of cultural control is, however, sometimes best seen where management practices change for agronomic reasons, and new pest problems are created as a result. Some examples of this will be given in the following discussion of cultural controls.

7.1.1 Soil cultivation

Many insects live or hibernate in suitable temperature and humidity conditions relatively near the soil surface. These conditions can be disturbed by ploughing, which creates temporary drought conditions in the upper soil layers and may even expose larvae and pupae to the full radiation of the sun. Many of these insects will be eaten by birds, and pigs have even been brought on to ploughed fields for the special purpose of picking out and eating white-grubs (beetle larvae pests of cereals). Other pupae and eggs may be buried by ploughing to a depth from which they fail to reach the soil surface after emerging. Then other individuals will be killed mechanically by rough contact with soil clumps; and root aphids (e.g. cereal root aphids) will suffer from the break-up of the ant colonies which tend them. The increased use of minimum tillage (whereby ploughing is given up in favour of the use of herbicides to kill any weeds on the soil surface) followed by the direct drilling of cereals into slots cut in the otherwise undisturbed soil surface has created numerous pest problems. Some of these relate to the earlier drilling of cereals permitted by this method, and will be discussed in Section 7.1.8. However, slugs and cutworms have benefited particularly from the lack of soil disturbance in minimum tillage systems, the mat of (albeit dying) weeds on the field surface providing suitable conditions in addition to the lack of disturbance of the soil.

Compacting the soil with a roller is a cultural measure for limiting the between-plant movement of some larger soil insects such as beetle larvae. Another problem with minimum tillage has been that of the stem-boring frit fly. This insect can develop large populations in grasses and migrate into winter wheat seedlings when wheat is drilled into a herbicide-treated sward in the common rotation of cereals after grass. This problem was never serious when the old sward was ploughed in late summer and left fallow until the sowing of spring wheat the following year.

7.1.2 Sanitation

Farm hygiene often has a pest control purpose. The destruction of crop residues removes residual pest populations (e.g. stalk-boring grubs in maize) and eliminates plant debris on the soil surface in which many pests find shelter for hibernation (e.g. flea beetles and whiteflies of brassicas). Destruction of crop residues of cotton followed by a gap before cotton is again planted is mandatory in many countries in the world, though not always enforced. It would be particularly effective against the bollworm *Platyhedra*, since this pest does not have any wild hosts to maintain the species if the crop population is destroyed. In banana and cocoa, pests often breed in fallen, rotting leaf or stalk material, and both fruit flies and the coffee berry borer often infest new fruit after emergence from fallen fruits on the plantation floor. The destruction of weeds acting as reservoirs for pest populations is often recommended, but is rarely practical. Thus it is recommended that nearby free-growing cotton and related weeds (Malvaceae) should be eliminated to reduce populations of the cotton-stainer bug in cotton fields. However, weeds outside the farmer's boundary are just as important as those within.

Another aspect of clean cultivation is 'roguing' – the removal and destruction of infected growing plant material where there is danger of spread to other parts of the crop. Before the advent of adequate plant resistance to the pest, the control of reversion virus spread by the blackcurrant gall mite was largely dependent on the removal and burning of infested bushes; roguing plants attacked by the sisal weevil are still a component of control of this pest in the tropics.

7.1.3 Manuring

The belief that vigorous plants are less attacked by pests is one of the foundation stones of so-called 'organic' farming, and it is far from being an erroneous concept. Rapid, healthy plant growth can reduce pest damage in four ways:

(a) Rapid growth shortens any susceptible stage. It therefore induces resistance against pests such as stem borers, to which seedlings have a relatively short window of susceptibility before the tissues harden;
(b) it may well lead to the maximum expression of some chemical resistance factors;
(c) it will allow maximum compensation for damage by the plant. For example, good root systems would clearly withstand root grazing by pests where weak root systems would not. Another example concerns the shot-hole borer (*Xyleborus fornicatus*) on tea in Sri Lanka, where damage was successfully reduced by fertilising the bushes with nitrogen. The stimulation in growth enabled the bush to form new tissue as a support bracket over the beetle gallery so that breaking of the branches at the gallery as tea pluckers passed through the plantation no longer occurred;
(d) it can promote uniformity and density of the crop stand. This can discourage pests such as the chinch bug (*Blissus leucopterus*), which is most abundant where the crop stand is somewhat thin. Aphids occur in smaller numbers where the crop is more dense; this is because fewer winged immigrants land where less bare ground is exposed.

However, just as fertiliser produces a more nutritious plant for man, so many insects may also benefit. Aphids, leafhoppers, mites, thrips and leaf-mining grubs have all been found to breed or develop more rapidly on plants given good nitrogen fertilisation. By contrast, there is some evidence that manuring with potassium and phosphate may reduce the incidence of some pests, and with aphids, which are sap feeders, good potassium fertilisation can reduce nitrogen available in the sap without impairing the value of the leaf protein. Here is a field ripe for useful and relatively simple experimentation – how far can we 'induce' plant resistance (see Section 6.7.4) by physiological treatments to plants?

7.1.4 Water and humidity management

Irrigation is a common practice in many crops, and it can be manipulated for pest control purposes. Small pests such as aphids are easily washed off plants by overhead irrigation, and soil insects may be killed by the pressure of swelling soil particles in saturated soils. Additionally, ample water availability causes physiological changes in plants; some sucking insects such as aphids and thrips tend to do badly on well-irrigated plants and benefit from periodic wilting of the plants. Raising the water level in rice paddies had been used to suffocate eggs of the armyworm *Spodoptera* and to drown larvae of the rice borer *Schoenobius*. Active insects such as shield bugs may be driven off the rice into the water, and subsistence farmers in Asia have been known to then put ducks onto the paddy to eat the swimming pests.

In the past, wine growers threatened by the dreaded *Phylloxera* took the desperate step of flooding their entire vineyard for long periods to suffocate the pest, in spite of the damage such flooding did to both the plants and the soil.

Where irrigation is essential to growing a crop, as in naturally arid areas in California and large areas of the American Middle West, the only lush vegetation in the region (i.e. the crop) may act as a magnet for pests. It is a well-known phenomenon in arid districts that pest incidence on cotton rises dramatically following irrigation. In California and Peru, however, irrigation has also enabled a pest/natural enemy complex to persist and reduce the importance of bollworms as pests.

Another approach to retaining moisture in the soil is to cover it with a mulch, often composed of plant debris. In coffee, thrips are rarely a problem in the more humid conditions of mulched plantations; just one season without mulch may elevate this insect to pest status. Damp conditions created by mulching may also be favourable to insect parasitoids; thus mulching in coffee increases the biological control of the *Antestia* bug. Still with coffee, pruning management is a further weapon against *Antestia*. This bug does less well where humidity in the canopy is reduced by pruning. Lower humidity, unfortunately, also makes the environment less suitable for parasites of the pest, but this can be compensated for by leaving the prunings on the ground as a mulch, and the reduction of biological control is minimised by careful timing of the pruning operation.

7.1.5 Strip farming and intercropping

Before intensive agriculture, farmers tended to grow several crops on one unit of land. Such multiple cropping is still common in peasant agriculture in the tropics. Either the area is divided into relatively narrow strips of different crops, or low crops are grown either under or in between the rows of taller crops (intercropping) (see Fig. 7.1). In intercropping, the low crop reduces weed competition by covering the ground rapidly, and prevents soil erosion and water loss.

Both strip farming and intercropping often reduce pest attack. In strip farming, the intervening strips of a non-suitable food may prevent movement of

Fig. 7.1 An intercrop of legumes between maize in Tanzania.

pests from one strip of a crop to another or from one suitable crop to a different one. Moreover, where two crops harbour unspecialised natural enemies, these can move over on to a neighbouring strip if pests build up there. The abandonment of strip farming in Peru some 40 years ago has been cited as one reason for the bollworm outbreak on cotton there, and certainly rediversifying the cotton agro-ecosystem there (Fig. 7.2) greatly reduced the incidence of the pest. The choice of adjacent crops is, of course, more important than the simple decision to diversify. Juxtaposition of wheat and maize, for example, would actually intensify the problems of shared pests such as chinch bugs and eelworms, whereas separating the crops by a strip planted with potatoes would reduce pest damage on the cereals. Strip cropping is of course unacceptable in highly

Fig. 7.2   Mixed cultivation in the Cañete Valley of Peru: part of a classic integrated control programme against cotton pests. Photograph courtesy of *The Furrow*, John Deere.

mechanised farming; moreover, some pests (e.g. some grasshoppers) lay eggs at the edges of crops and can become a serious problem when, as with strip farming, the edge forms a large proportion of the crop.

Intercropping seems to have three main effects on insects which result in lower pest numbers:

a)   Intercropping may reduce pest damage by attracting the pests to a less valuable crop, or one where the pest is less serious for some reason. One example is the mixing of maize and cotton to achieve control of the shared lepidopteran pest *Heliothis*. This tactic can backfire if not timed perfectly; the formation of young bolls on the cotton must coincide with the tasselling of the maize. *Heliothis* is attracted to the maize tassels, but is a much less serious pest on maize than cotton, because attack on the cobs is reduced by cannibalism when larvae meet within the tight husks. Intercropping cowpea with sorghum can attract polyphagous pests onto the sorghum, which is a less valuable crop.

b)   The host-plant-finding behaviour of insects may be disrupted by the close juxtaposition of two plant species. Several crop pests, such as cabbage white butterflies and cabbage aphids are very much influenced by the crop background in their colonisation behaviour, and intercropping removes the contrast between seedlings and bare soil in the same way as dense planting does. Weeds, of course, have the same effect as a low intercrop; it has been shown that very few immigrant aphids were trapped over weedy plots of Brussels sprouts. Additionally, the mixture of odours from an intercrop, particularly any strong smell from a non-host plant for the pest masking the odour of the host plant, can disrupt the host-finding behaviour of pests. This has been shown, for example, by work at Cambridge in relation to carrot fly on carrots interplanted with onions. There are many other, mainly anecdotal, records of aromatic plants repelling insect pests, particularly those of vegetable crops, and this is another area which is very amenable to a little experimentation.

c) Intercropping may also increase the impact of natural enemies. This may be because one of the intercropped plants provides a honey or nectar source which attracts natural enemies for adult feeding, or because the shelter and humid conditions near the ground provided by the intercrop encourage ground-living predators. Work at Rothamsted Experimental Station has shown that many predatory beetles are more abundant in weedy rather than clean plots of winter wheat. There is evidence that some ladybirds prefer ground cover to rows of plants in bare ground. Some hover fly adults, whose larvae are important predators of aphids, also lay more eggs where the ground is covered than where it is bare; unfortunately other hover flies have the reverse behaviour. Another experiment at Rothamsted Experimental Station had endeavoured to establish aphid parasitoid populations by undersowing cereal crops with rye grass and liberating an aphid, which lives on the grass but does not attack wheat, together with its parasite. The parasite, however, also attacks grain aphids, and there is some evidence that this procedure improved biological control of the aphid. Work on cabbage root fly in plots of cabbages either grown traditionally or undersown with clover has shown that the clover undersowing greatly promotes the number of ground beetle predators of cabbage root fly eggs.

7.1.6 Crop rotation and isolation

Attempting to separate the pest from its host plant in time or space is one of the oldest and most widespread farm practices often directly motivated by pest control, and it is still one of the most effective controls of some eelworm problems. Crop rotation normally reduces and delays attack rather than giving complete control because, although control may be significant within a given field, it is a less effective restraint over an area as a whole. Most pests have strong migratory powers or, if not, can frequently survive rotation on wild host plants (see Section 7.2.1). Moreover, crop rotation usually means that a particular crop is nevertheless grown somewhere close-by in the area. Thus the common rotation for a field of grasses or cereals, followed by legumes and then root crops does not result in the absence of any of these crops on the farm as a whole. Yet the rotation is effective in reducing the many soil pests (e.g. wireworms, chafers, leatherjackets and shoot-boring flies) which multiply most successfully under grass. The various crop midges (e.g. pea midge and bladder pod midge) are weak fliers and also are affected by crop rotation. However, just to emphasise the point that cultural controls can often be a two-edged sword, it is worth giving the example of the wheat bulb fly (*Leptohylemyia coarctata*), which strangely does not lay eggs in wheat crops, but in any fallow ground. The pest is therefore not a problem when cereals follow cereals, but only when cereals follow, through a rotation, bare fallow or a crop such as a root crop which leaves a fair amount of the soil surface exposed in late summer.

Crop rotation relies on the fact that there are usually only a few general feeders among the pests found across the rotation. For example, of 50 serious insect pests of the maize, wheat and red clover rotation, only three are important pests of all three crops.

Attempts to avoid pests by isolating crops from regularly infested sites are frequently designed to prevent insect-borne diseases from reaching the isolated crop. Because wild plants (see Section 7.2.1) form reservoirs of both the insect vectors and the diseases they carry, the method has rarely proved successful on a regional scale.

7.1.7 Trap crops

If insect pests can be concentrated in particular small areas of a field, they can then be destroyed with locally applied pesticide or some other technique to which insects are unlikely to develop resistance, such as ploughing in, feeding the vegetation to livestock, or the use of a flame gun. Such concentrations of pests may be induced by position (e.g. edge rows for swede-midge); by exploiting the crop zone in which most insects are deposited behind windbreaks (see Section 7.2.2); by planting taller plants at the edge of the crop to filter out flying insects; by earlier sowing (e.g. against the corn earworm); and by spraying with attractants or choosing especially attractive plants as the trap crop (e.g. kale for certain bug pests of cabbages). A particularly ingenious example of a trap crop is the use in Canada some 30 years ago of a non-crop trap plant (brome grass) planted in a 15- to 20-m strip around wheat fields to control the stem-boring sawfly *Cephus cinctus*. The adults did not penetrate into the wheat crop but laid eggs on the brome grass in which many larvae developed per stem. It was not necessary or even advisable to destroy the grass, for the grubs cannibalised one another, and even most of the eventual survivors failed to survive to maturity in the grass, although their parasites were able to emerge. Thus the brome grass filtered out the sawflies and effectively converted them into biological control agents for the crop!

Trap crops are therefore either non-crop plants or crop plants intended to be especially heavily damaged by pests and often destroyed well before harvest. The land occupied, which in some instances may need to be in the order of 5 to 10% of the whole crop, inevitably involves some reduction in crop yield.

7.1.8 Sowing and harvesting practices

Variation of sowing date can control pests, most of which show some seasonal predictability, either by avoiding the egg-laying period of the pest or by allowing the plants to reach an age where they are resistant by the time the pest appears. For example, the hessian fly (*Mayetiola destructor*), has a predictable flight peak of limited duration; thus a few days delay in sowing wheat can make all the difference between a good and a bad crop. Changes in the sowing date for agronomic reasons have sometimes caused new pest problems. The increased growing of winter wheat, sown even earlier to exploit non-tillage systems (see Section 7.1.1), has created a whole range of new problems related to the changed timing of the crop. A stem-boring fly (*Opomyza florum*) now finds cereals at the right stage of growth when it lays its eggs in late summer. Also, the winter cereals are infested by the aphid *Rhopalosiphum padi* in autumn and can be infected with barley yellow dwarf virus by this aphid. The virus then multiplies in the plants for the rest of the crop season and can cause severe symptoms the following year. Spring wheat avoids this autumn aphid infestation and, although the virus may be brought in by other aphids much closer to harvest, the virus is then not much of a problem.

Seed rate can also have a considerable effect on pest problems. It has already been mentioned that fewer insects exist in dense rather then sparse stands (Sections 7.1.3 and 7.1.5). Many crops, such as cereals, are sown densely, and yield per hectare tends to remain constant over a wide range of sowing densities because of interplant competition. The loss of some plants due to pest attack is therefore relatively unimportant. Sugar beet seed used to be polygerm (i.e. each seed produces several seedlings), and was sown thickly and later thinned to stand. Recent changes in crop management have included precision drilling to plant stands of wheat and sugar beet, made possible with the latter by the development of monogerm seed. Attack on seedlings of these crops is now much more critical, and insects such as pygmy beetle and mangold fly, which used to be regarded as unimportant pests of sugar beet, have now become far more severe.

Early harvesting may remove pests (especially cereal pests in the straw and grains) from the field before they can emerge and perpetuate the population in the area. Damage to wheat caused by the wheat stem sawfly can be minimised by harvesting early, before the weakened stems lodge in wind and rain.

7.1.9 Conclusions on cultural control

Although it is often possible to conceive of a cultural control method which could cause considerable reductions in pest damage, it is much harder to think of an approach which would also be acceptable to the farmer, particularly in developed agriculture (but see Section 7.2.3). However, agronomists and cropping systems scientists in many institutions are continually setting up experiments of new cultural systems which they believe may gain farmer acceptance for economic and agronomic reasons; it may well be just as sensible for the worker interested in pest control to take data from the experiments of these colleagues than to go it alone and try to develop new systems purely for pest control purposes. At the very least, by now history should have taught us that proposed changes in crop management can cause new pest explosions.

7.2 The importance of non-crop plants in crop pest problems

Comments in the previous section about destruction of weed hosts, about the problems that alternative plant hosts present for crop rotation and about strip cropping or intercropping indicate that other plants, particularly 'wild' flora, in the agro-ecosystem can be of considerable importance. The subject has been reviewed by Lewis (1965*a*) and van Emden (1965, 1981). It is possible to distinguish biological and physical components in the relationships of pests and natural enemies to non-crop plants. Biological components are those where particular species of wild plants are used for feeding by insects. Physical components are largely microclimatic; the species composition of wild flora is relatively unimportant, and feeding by the insects is not involved. In Britain, hedgerows, roadside verges and patches of woodland are important areas of non-crop plants in the agricultural landscape; many biological components also relate to weeds present in the fields themselves.

7.2.1 Biological components (see Fig 7.3)

Most farmers realise that wild plants can act as a reservoir of crop pests, especially if the wild plant is botanically related (*1a*) to the crop. Thus there are may insect links between grasses (particularly ubiquitous in arable areas) and cereals, between cotton and wild Malvaceae, and between potatoes and wild Solanaceae (which are common hedgerow roadside plants). However, there are also numerous examples where alternative wild hosts of crop pests are totally unrelated to the crop (*1b*); for example, a weevil pest of strawberries (Rosaceae) also feeds on stinging nettles (Urticaceae).

Pests may use wild plants as food if the crop season is shorter than the insect feeding season (*1c*). Also the hardening of tissues on early weeds or irregular events such as the use of weedkillers on roadside verges may 'drive' insects on to crops (*1c*). Wild plants can therefore form a reservoir of crop pests which attack the new crop in the spring; they can also maintain the species in the area if the crop is absent because of rotation. In addition to forming a pest reservoir, non-crop plants also form an important reservoir of crop diseases from which the new crop can be infected each year. Barley yellow dwarf virus of cereals, for example, has a permanent reservoir in grasses. Very often infected wild plants show no symptoms of disease, and its presence can only be shown by electron microscopy, immunological techniques such as the ELISA test or the infection of crop plants when the vectoring insect is transferred. Various impressive

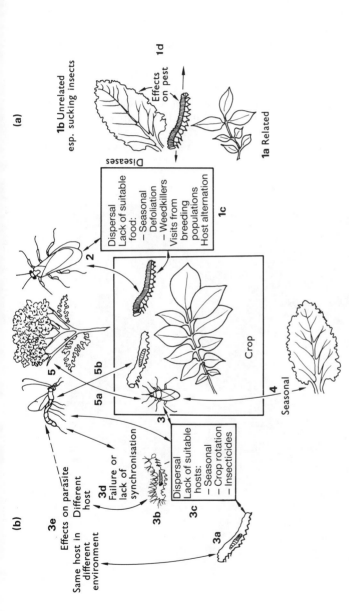

Fig. 7.3 Biological relationships between non-crop plants and both pest (a) and beneficial (b) insects. Injurious species feed on both crop and related (1a) or unrelated (1b) wild plants for a variety of reasons (1c). Pest insects (2) predators, (5a) and parasites (5) may also feed on flowers. Beneficial insects may utilise alternative prey related (3a) or unrelated (3b) to the crop pests, for several reasons (3 c, d). Alternative food may affect the physiology of pests (1d) or benefical insects (3e). Sometimes carnivores may also feed phytophagously on non-crop plants (4). From van Emden (1965), courtesy of the late Horticultural Education Association.

examples can be found of virus diseases becoming evident in crops as soon as the crops are introduced into new agricultural areas.

The different nutrition that different plant species provide can affect pests arriving on crops from outside vegetation in various important ways. They may, for example, show a higher fertility and a greater tolerance to insecticides and insect pathogens than insects reared on the crop itself (*1d*).

Flowers are very important sources of food for adult insects (Fig. 7.4), and the abundance and diversity of insects visiting patches of wild flowers, particularly plants like hedge parsley in the family Umbelliferae, is often exploited by insect collectors! Many insects have to feed on pollen and nectar as adults before their eggs can mature. Several pest insects (Fig. 7.3) (*2*) fall into this category, and it has been calculated that eight cow parsley plants produce enough nectar to feed at least 2000 cabbage root fly adults between emergence and oviposition. However, the beneficial impact of flowers in maintaining parasites and predators of crop pests (*5*) usually far outweighs their possible detrimental influence (see also biological control, Section 3.4.3). The routine use of herbicides in modern agriculture means that flowering weeds outside the crop are frequently the only source of flowers for beneficial insects in the agro-ecosystem.

Fig. 7.4   A solitary parasitic wasp of insects (*Ammophila* sp.) feeding at a flower.

For the same reasons that pests utilise plants outside the crop for food, the various, often non-economic plant-feeding insects such weeds support may be valuable or even essential for maintaining natural enemies in an area (Fig. 7.3, *3b*) (see Section 3.4.3 for fuller discussion). When land is left fallow, or when insecticides are used intensively, prey on uncultivated land may be essential to perpetuate natural enemies in the locality (*3c*). Some general predators (e.g. some bugs) may even turn to plant feeding when animal prey is scarce (*4*). As with pest insects, so natural enemies reared outside the crop may differ in biological characteristics from those reared in the crop (*3e*). There is evidence, for example, that some parasites reared on prey on weeds may seek that same weed to find hosts for their next generation rather than seeking the crop. Relatively little is known about the effects on beneficial insects of feeding on prey on weeds, yet these matters are amenable to quite simple experiments.

7.2.2 Physical components

The interference to air currents caused by an upstanding barrier (e.g. a hedge) causes turbulence on the lee side, and small airborne insects (many of which are pests) are deposited on the crop by the down currents so caused. Some deposition of insects also occurs close to the windward side of the hedge (see Fig. 7.5, 1).

The area of deposition is determined by the height and permeability of the barrier, and Lewis (1965*b*) demonstrated these effects brilliantly with lettuce and an artificial windbreak – the root-aphid-infested lettuce died and the dead

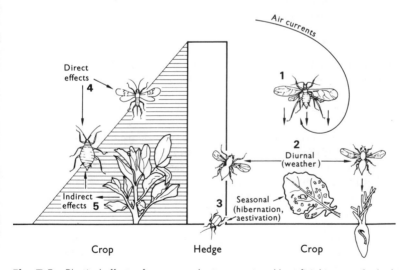

Fig. 7.5 Physical effects of non-crop plants on pest and beneficial insects. Such plants may interfere with the transport of insects in air-currents (1) or provide diurnal (2) or seasonal (3) shelter. The microclimatic conditions adjacent to a hedge may influence the activity of insects (4) or indirectly affect them (5) via effects on the host plant.

Fig. 7.6　Damage to lettuce by the lettuce root aphid (*Pemphigus bursarius*) illustrating the localised deposition of the immigrating aphids in the shelter of a 45-per-cent-permeable fence. During the immigration, the prevailing wind was approximately right to left across the picture. The smaller windward and larger leeward areas of shelter are shown by the bare areas where lettuce was killed out by aphids. Photograph courtesy of Dr. T. Lewis.

plants stood out as brown bands showing the exact area of deposition (Fig. 7.6). The same pattern can be seen in the deposition of snow behind a hedge or fence, and deposition of disease-carrying aphids accounts for the 'bands' of yellow virus-infected beet and potatoes frequently noticed near the margins of fields.

A different pattern of infestation, where crop damage falls away directly from the edge of the field towards the centre, is characteristic of many of the root fly pests, such as cabbage root fly and carrot fly (Fig. 7.5, 2). The adults 'roost' at night or during adverse weather and invade the crop merely to lay their eggs. Other insects seek the shelter of debris and fallen leaves in the hedge (Fig. 7.5, 3) for overwintering (e.g. flea beetles) or aestivation (e.g. apple blossom weevil).

A further physical impact hedges may have is through the shade and shelter they give over part of the crop area. The effect may be direct – for example, most insect parasites fly and seek prey more actively in the sun than in the shade (Fig. 7.5, 4) – or indirect – in that plants which are shaded and suffer competition for moisture and nutrients with adjacent bushes and trees usually provide poorer nutrition for pests than those in the open (Fig. 7.5, 5).

7.2.3 Conclusions on non-crop plants

Non-crop plants clearly have both debit and credit properties with respect to pest control. The accumulation of insects around windbreaks is on balance harmful, and the practice of improving road vision for motorists by cutting roadside verges tends to remove the beneficial element of flowers while retaining the pest reservoir of herbivores on the foliage. However, non-crop plants do have certain essential irreplaceable properties, particularly in providing sources of adult food for and maintaining a reservoir of natural enemies. There are also probably many examples to be discovered similar to that of *Angitia* parasitising *Plutella* (see Section 3.4.3) waiting for man to break the important trophic link.

Intensification of farming in Britain has of course involved the removal of hedgerows in many areas, such as East Anglia, and the increasing use of herbicides. The reduction in abundance of wild plants in agricultural areas has been noticeable, but there is no equivalent sign that the type and severity of pest problems experienced by farmers has simultaneously changed. If wild plants were on balance harmful in pest terms, these changes in the landscape should have been accompanied by a decline in the need to use insecticide. Equally, on the other hand, there has been no increase in the use of insecticides attributable to the loss of the beneficial effects of wild plants. Probably wild plants are still sufficiently abundant, and the flight range of insects so large, that it is still impossible to guess at the extent and distribution of wild flora which have to be retained to maintain the biological control impact of wild plants in the agro-ecosystem. It is, however, obvious that we must retain some 'weed' outside the crop if we require any contribution from natural enemies in our pest control. Thus, although a farmer's own wild plants are probably a nuisance to him in pest terms, it does seem essential that other farmers keep theirs.

Recently, and particularly in relation to cereal crops, workers at the Game Conservancy and at Southampton University have made some progress in reconciling the need for plant diversity in crops with the economic aspirations of farming. The income the farmer can obtain by making the edges of his fields more productive of insects to achieve better survival of partridge chicks has encouraged some farmers not to spray herbicides on broad-leaved weeds or insecticides on a 6-m-edge strip of their fields. Also, the exact characteristics of hedge banks which encourage predatory ground beetles have been identified, and researchers are now experimenting with 'mini-hedgerows' that take up no more space than post and wire fencing as well as with types in the form of narrow specially designed and sown strips criss-crossing cereal fields. Economic evaluation suggests that the cost of insecticide that enhanced biological control may achieve more than compensates for the small yield loss of devoting a small fraction of the field to these little hedgerow facsimiles.

7.3 Legislative control

This is the legal enforcement of certain control measures or inspection procedures.

7.3.1 Quarantine

Most countries operate quarantine laws to allow inspection at the point of entry of all produce which might harbour foreign pests; these laws also enforce strict isolation of any species imported for study (e.g. for biological control research). Unfortunately, quarantine normally only postpones the entry of pests, and such entries have become more frequent since the advent of mass air travel. Perhaps propaganda, so that travellers and importers are more aware of the dangers of introducing new pests, is one of the most important components of the quarantine system. Strict and largely successful quarantine as practised, for example, by Australia, does mean that crop varieties are bred in the absence of some of the most important world pest problems. When such pests eventually arrive, they can cause devastating damage, as the plant breeder will not have checked his high-yielding varieties for susceptibility to these organisms.

7.3.2 Eradication

Particularly serious pests may be subject to a 'Notification Order', whereby any farmer who suspects that the pest may have appeared on his crop must notify the appropriate authorities, who then undertake pest eradication. The Colorado beetle in Britain falls into this category.

7.3.3 Certification

Certain plants, seeds, tubers, etc., subject to particular pests or diseases may not be sold unless free of the problems. For example, 'The Sale of Diseased Plants Order' (a series of laws passed in Britain between 1927 and 1952) prohibited the sale of plants with infestations of several pests. These pests included glasshouse whitefly, but 'The Sale of Diseased Plants Order' was probably more honoured in the breach than in the observance.

7.3.4 Rotation orders

Rotation is among the cultural practices which have been subject to legal enforcement in various countries at various times (e.g. sugar beet rotation to control beet eelworm in Britain).

8
Other control methods

8.1 Introduction

The problems of the continued widespread use of pesticides, and particularly
the absorption of the words 'environmental pollution' into common vocabu-
lary, have caused scientists to look seriously at any ideas for pest control which
do not involve traditional insecticides.

Although chemical, biological and cultural control and the use of resistant
plant varieties are the four pest control methods which have the widest general
application, there are quite a lot of other methods which are, or have been, in
use or have been proposed.

8.2 Physical controls

Such controls aim to reduce pest populations by using devices which affect
them physically or alter their physical environment. These may be hardly dis-
tinguishable from cultural controls and are frequently labour-intensive. For
example, in the early days of pest control in developing countries, hand-pick-
ing and foot-crushing of larger pests (e.g. caterpillars) was economically viable
and effective. Grease bands around the trunks of apple trees to trap the ascend-
ing flightless females of winter moth and tarred discs around cabbage plants to
prevent cabbage root flies from laying eggs close to plants were standard
practice for many years before the advent of modern insecticides.

Most such laborious practices have now proved too expensive, and rather
sophisticated machinery now represents 'physical controls', though as yet most
of it is still on the research bench. Here we can cite as examples traps with fans
to disintegrate entering pests and the amplification of attractive or repellent
sounds, respectively, to attract pests to some other doom (e.g. insecticides) or to
drive them from the crops (see Section 8.3).

For a short period in the late 1950s, sticky sprays (polybutenes) were applied
to plants to trap small insects and mites when they emerged from the eggs, yet
not to trap larger beneficial insects. Plants often reacted rather badly to these
sticky films, and they are no longer used.

The only method in this category which has really stood the test of time is hot water treatment of plant storage organs (e.g. roots, corms and bulbs) to kill concealed pests such as bulb flies and eelworms. Unfortunately, generality of the technique is limited because the thermal death point of many pests is quite close to temperatures which damage the plant organ.

8.3 Repellents and attractants

8.3.1 Repellents

Chemical repellents have been investigated most frequently with respect to mammals and birds, which are often difficult to control by other means. However, it has been hard to find chemicals which do more than influence choice by the animals; moreover, the compounds concerned are often not very persistent and require frequent reapplication. For a long time various 'folk remedies' have been used to repel pests with herbs and other plants, and oil of citronella has been claimed as particularly effective. The most widely used insect repellents have been those against mosquitoes and other biting flies attacking human beings.

Repellent 'signals' of other kinds have also been explored, particularly repellent sounds. Here there has been most interest in the avoidance action taken by certain moths when pursued by bats, and this has led to the production of amplified ultrasonic sounds in crops to mimic bat calls and drive moths from orchards and cereals.

Other work on repellence has exploited the fact that short wavelength light (sky) reflected from pieces of aluminium foil laid between the rows in a crop can greatly reduce the number of aphids landing on the plant by inducing descending individuals to attempt to fly upwards again. The material and labour costs of this method are high, but it has been used commercially in some high-value crops (e.g. in the cut flower industry), where it has given effective control of aphid-borne virus disease.

8.3.2 Attractants

The use of attractants against insects has been developed much more than the use of repellents. Attractants can be combine with other control methods to introduce selectivity; i.e., the species to be controlled is selectively 'lured' to its doom! Among chemical attractants, the high specificity of sex pheromones (see Section 5.2) has proved of particular value. Attractant chemicals have also played an important role in the control of fruit flies. These insects are attracted by the decomposition products of fruit, and lures developed from these have been used effectively to produce attractive baits treated with insecticides and distributed within (or better still outside) the crop area.

It has long been known that light is attractive to many insects, especially those flying at night. Light traps have never given very high levels of insect control, but one device commonly used in public health is an electric-grid light trap which electrocutes the attracted insects. There have always also been experiments with broadcasting mating calls, particularly those of grasshoppers, from traps.

8.3.3 Antifeedants

The distinction between repelling an insect and inhibiting its feeding is that in the latter case the insect remains on the treated plant and starves to death rather than dispersing to seek food elsewhere. Some food for natural enemies therefore remains for a time. Antifeedants appear to act on the taste receptors of the insect and inhibit their perception of the stimuli to feed present in the host plant. Antifeedant properties are common in chemicals used for other crop protection purposes, for example in triphenyl acetate (fungicide), some triazenes (herbicides) and carbamates (insecticides). Carbamates may show antifeedant action at rates well below those lethal for the insect. Knowing that triphenyl acetate had antifeedant action, researchers began work on organotin compounds in general, and a number of new compounds (e.g. triphenyltin) were shown to be particularly promising. As yet, no systemic antifeedants have been discovered, so sucking insects which pierce the treated surface are not affected, and neither is there any protection of new growth or of leaf area missed by poor coverage.

8.4 Genetical control

Genetical control interferes with the ability of insects to reproduce without killing them. The 'treatment' of one individual which is then released alive may have a much greater influence on the size of subsequent generations than killing it outright; this is known as the 'one to many' principle.

8.4.1 Radiation sterilisation

The control of the screw-worm fly (*Callitroga hominovorax*) of cattle by radiation sterilisation (Knipling, 1955) is a landmark in the history of pest control. Knipling considered *Callitroga* an ideal target, since it fulfilled the theoretical requirements he saw as essential for best use of the technique:

a) a method for the mass rearing of males;
b) the released males must disperse rapidly throughout the native population;
c) sterilisation must not affect sexual competitiveness;
d) preferably the females only mate once (e.g. *Callitroga*).

He also developed a model (Table 8.1) on how the population of the pest might decline in a sequence of generations with a constant number of sterile

Table 8.1   Theoretical population decline in each subsequent generation when a constant number of sterile males are released among a natural population of 1 million females and 1 million males. From Knipling (1955).

Generation	Number of virgin females in the area	Number of sterile males released each generation	Ratio of sterile to fertile males competing for each virgin female	Percentage of females mated to sterile males	Theoretical population of fertile females each subsequent generation
F_1	1 000 000	2 000 000	2 : 1	66.7	333 333
F_2	333 333	2 000 000	6 : 1	85.7	47 619
F_3	47 619	2 000 000	42 : 1	97.7	1 107
F_4	1 107	2 000 000	1807 : 1	99.95	Less than 1

males released per generation. It is clearly important to swamp the natural population with sterile males, and therefore Knipling suggested that sterile males should be released after the native population had been reduced by insecticides. This has the nice touch of using insecticides in their most effective way of bringing down high populations, and then using sterile males in their most effective way, when the population is low. In experiments on a Pacific island, where there was of course a finite fly population without immigration, the release of males sterilised by a 5000-Röntgen unit cobalt bomb resulted in an eradication of the fly in 8 weeks. The technique was then transferred to the south-eastern United States, where again there is a relatively isolated population of the pest. At the height of the campaign in 1958, more than 50 million flies were reared, sterilised and released every week. The fly was virtually eliminated, and the continual invasion of wild flies from across the Mexican border was countered by the frequent release of sterile males in this area. This was accomplished by the erection of a factory on assembly line principles on a disused airfield. From an input of more than 50 tons of blood and meat a week, the factory produced 150 million sterilised pupae for airlifting into the countryside in parachuted cardboard containers. However, over 90 000 cases of screw-worm were again reported in 1972, and it became evident that continual factory rearing had led to the sterilised males no longer being readily accepted by the wild females. As a result, the factory stock is periodically revitalised by the introduction of new wild material.

The technique has also been attempted against other insects. Some success was obtained in Louisiana in 1962 against the sugar-cane borer, by releasing sterile males at a time when the wild population was at a naturally low level. However, most work has centred on the tropical fruit flies (the family Trypetidae), and these have been controlled in various Pacific islands by the sterile insect release method (the acronym SIRM is now commonly seen in the literature), although the females mate several times. There is a long-standing research programme at the Atomic Agency Authority in Vienna developing fruit fly control by SIRM.

8.4.2 Radiation-induced translocations

Research on this technique has concentrated on the control of mosquitoes, but there seems no reason why it should not be applied to other pests. The technique has two advantages over radiation sterilisation: both irradiated males and females can be used, and the release of sterile individuals needs to be only half rather than double that of the wild population. The aim of the technique is to cause exchange of chromatin between chromosomes (translocation) by radiation and then to mate these individuals to a homozygous form (Fig. 8.1). When these homozygotes are released and mate with wild individuals, half the off-spring receive a lethal combination of gametes, a quarter are viable but carry the translocation and the remaining quarter are normal viable insects. Thus, not only is the potential population halved, but a half of the viable population carries the translocation to hand on the lethality to subsequent generations. This contrasts with radiation sterilisation, where the released males die out

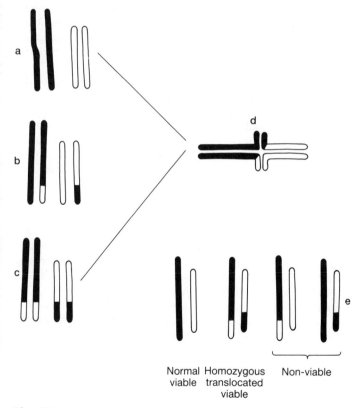

Normal Homozygous   Non-viable
viable translocated
        viable

Fig. 8.1 Radiation-induced translocations. (a) Normal chromosomes; (b) chromo-somes with induced translocation (c) translocated homozygote (d) pairing of translocated homozygote with normal homozygote; (e) gamete possibilities.

each generation and have to be replaced by other bred and sterilised individuals.

8.4.3 Chemosterilants

In many ways, sterilisation can be achieved more simply with chemicals than by exploding cobalt bombs or using X-ray sources, and there has been considerable interest in such chemicals because they could be manufactured and sold by the agrochemical industry. Most of the effective chemosterilants have proved to be either anti-metabolites (e.g. fluoracil) which act as substrates competing for enzymes in nucleic acid synthesis, or alkylating agents (e.g. apholate, tepa) which replace hydrogen atoms with an alkyl group, again particularly in nucleic acid synthesis. The alkylating agents have been particularly successful in laboratory experiments with a wide range of pests. However, the alkylating agents are also mutagenic for man. There thus seems little prospect of them being applied directly to crops or other substrates in the same way as pesticides, and most trials with chemosterilants have involved traps baited with an appropriate pest attractant (often a sex pheromone). Most chemosterilants seem particularly effective in male insects, which is perhaps an advantage, since the sex pheromones which can be used to attract pests to the traps also are largely male specific.

There are, however, examples (e.g. amethopterine against house flies or the fly *Drosophila*) where the females appear the more susceptible sex.

Sterilising agents can effect reproduction in treated insects in a number of ways, but the effects of potential practical use so far identified have been the production of lethal factors in the sperm. Thus when glasshouse red spider mites are treated with apholate, the male is sterilised and matings result in the production of male offspring and dead eggs (representing the female fraction).

8.4.4 Hybrid sterility

Geographical races of the same insect may not always be genetically compatible. Particularly in Lepidoptera, the intensity of male-determining factors seems to vary geographically. Liberating large numbers of males from a foreign strong race would act like sterile male release, in this case producing large numbers of normal males but sterile intersex females.

With other insects (much research has been done with mosquitoes) the hybrid males of a cross between geographical races may be viable but sterile, and could theoretically be used for sterile-male release campaigns.

8.4.5 Competitive displacement

This is the concept of introducing a highly successful competitive species to displace and replace an existing species in an area. A plant pest would therefore usually need to be replaced by another plant feeder, and at first sight nothing would appear to have been gained. However, there are quite a number of advantageous possibilities which include:

a) the newcomer need not be a potential pest if it out-competes the resident species in some arena other than competition for food from the crop (e.g. on wild plants in the fallow season, or for pupation sites);
b) the newcomer, although a potential pest, is at least not a disease vector as the resident is, and a plant disease problem may therefore be more easily controlled;
c) the newcomer may be severely affected by adverse climate (e.g. it is non-diapausing in temperate conditions). The release of the newcomer, if it were able to displace the resident in one season, would then end in the disappearance of both species.

An example in the last category concerns the field cricket (*Teleogryllus commodus*) in Australia. The race of the cricket in the colder south produces diapausing eggs, but the race in the north does not. When southern females are mated to northern males they produce only non-diapausing eggs, and thus matings following the release of northern males into the south should result in many eggs laid by the southern females failing to survive the winter.

One problem with the concept of competitive displacement is that there is usually no reason why the competitor should out-compete the resident species. Especially where the introduced competitor would need to replace the resident species within one season, it may be necessary to give the introduced competitor some selective advantage. An obvious technique would be to breed insecticide resistance into the competitor, and then manipulate insecticide use to exploit this artificial advantage.

8.5 Internal growth regulators

These are chemicals which interfere in an adverse manner with the normal growth and development of insects (Fig. 8.2). The idea of using such products originated from an insect endocrinologist, Carroll Williams, who suggested that the hormones produced internally by insects to regulate their moulting and metamorphosis could be turned back on them as 'third-generation insecticides'. Original claims made for such third-generation insecticides were that they should be highly specific, and that resistance to them would be unlikely.

Insect hormones are relatively non-specific among insects, but certainly they are likely to be less damaging to other types of organisms in the environment than are insecticides.

However, resistance to hormones and 'hormone mimics' has occurred, presumably because insects already possess the compounds to maintain their own hormone titre internally.

8.5.1 Juvenile hormone analogues

These function in the same way as the juvenile hormone in the regulation of metamorphosis, and may or may not be similar in chemical structure to the

Fig. 8.2   Pupal deformity following application to large cabbage white butterfly (*Pieris brassicae*) caterpillars of a chemical with structural similarities to insect moulting hormone. Photograph courtesy of ICI Agrochemicals.

natural hormone. Two such analogues are methoprene and kinoprene. The effects of these compounds are usually seen during larval to pupal metamorphosis, and various degrees of incomplete metamorphosis become apparent. Larval-pupal mosaics may be produced, or strange deformations may appear on the pupal structure. Other uses of juvenile hormone analogues are in disrupting embryogenesis in the eggs and in preventing adult diapause. They have been extensively tested in public health and stored products work because of their relative safety to human beings.

8.5.2 Anti-juvenile hormones

A high titre of juvenile hormone in the insect maintains its larval characteristics, and if this effect can be counteracted in the early stages, then the larvae may metamorphose into miniature pupae or sterile adults. This is known as precocious development, and the name 'precocenes' has been given to a major group of anti-juvenile hormone compounds. None of the compounds developed so far have been sufficiently active for practical purposes, but there is still hope for the future.

8.5.3 Chitin synthesis inhibitors

The benzoylphenyl ureas, which interfere in chitin synthesis in insects, were first discovered about 1970. Since then a number have been marketed quite successfully, particularly against the Lepidoptera. These compounds interrupt the organism's moulting process. Although the new skin seems to be formed normally, it is the shedding process which is disrupted, and affected insects either die within their old cuticle or fail to emerge satisfactorily from it. Though effective, the compounds are still rather expensive compared with traditional pesticides. Nevertheless, chitin synthesis inhibtors are particularly useful where some selectivity of action is required (parasitism is usually not reduced by application) or where the pest has become resistant to insecticides.

9

Pest management

9.1 Introduction

As was mentioned in Chapter 1, man has already found situations where the insecticide 'road' ran out. The available insecticides failed in the middle and late 1950s because of tolerant pest strains on cotton in Peru, on lucerne in California and on chrysanthemums under glass in Britain. Rachel Carson (1962) had advocated that man must choose between chemical and biological control; he (man) was 'standing at a fork' of the ways. The first sentence of her final chapter ('The Other Road') in fact begins with the sentence 'We stand now where two roads diverge'. In the light of what was already happening at that time with lucerne in California, and what has happened since elsewhere, the sentence stands out as perhaps the most interestingly misled ever written about pest control.

It is a useful exercise to look at the above three famous examples of failure of insecticidal control in the 1950s and to consider the principles involved in the solution of these problems in the light of that sentence of Rachel Carson's, 'We stand now where two roads diverge'.

9.2 The classic examples of insecticide failure in the 1950s

9.2.1 Cotton pests in the Cañete Valley of Peru

The Cañete Valley was a vast irrigated cotton area in an otherwise dry and lifeless region. Ecologically it was an island of monoculture and therefore predisposed to ecological disaster once the area was regularly blanketed with insecticides. Since the entire ecosystem, including its faunal complement, was regularly treated with chemicals, there were rapid and devastating problems. New pest problems appeared and, together with pesticide resistance, caused a yield crisis as early as 1955. In 1956 a legislative package was introduced based on compulsory crop rotation or a mixture of crops on every farm (Fig. 7.2), a return to older insecticides (particularly lead arsenate, which is a stomach poison, and therefore the leaf surface is not toxic to beneficial insects) and the reintroduction of beneficial insects. The result of this package was that yields of cotton began to rise dramatically in the late 1950s.

9.2.2 Spotted alfalfa aphid in California

The spotted alfalfa aphid (*Therioaphis trifolii*) was first seen in California in 1954, and was presumed to have been introduced from Europe. By the late 1950s the aphid had developed resistance to organophosphate insecticides, and crop losses became critical. The courageous step was taken of applying an organophosphate (dimethoate), not at an increased dose, as is usual when resistance problems appear, but at a reduced dose. Some aphids were still killed; of course many survived, but so did many of the natural enemies which had not been effective controls on their own before but were now able to control the surviving aphids. The local natural enemies were also reinforced by the importation of three additional species of aphid parasites. Within a year of applying this programme, the crisis was over. As a cultural measure, strip harvesting of the lucerne was introduced, in order to maintain some aphids and natural enemies on newly cut strips when older ones needed some insecticide treatment. Later on, varieties resistant to the alfalfa aphid were introduced, but the early solution of the insecticide crisis with lucerne came before these resistant varieties were available.

9.2.3 Peach potato aphid on chrysanthemums under glass in Britain

The discovery of lighting treatments to produce flowering pot chrysanthemums at any time of the year led to the concept of all-year-round chrysanthemums as a crop permanently occupying large glasshouses in southern England. Thus the peach potato aphid (*Myzus persicae*) now bred continuously (Fig. 9.1), and resistance to organophosphate insecticides quickly developed. By the early 1960s, this resistance was 4000-fold. At first, the solution was thought to be biological control, but attempts to use aphid parasites proved unsuccessful. However, it was realised that, if broad-spectrum insecticides were to be avoided, another control for glasshouse whitefly, red spider mite and thrips would need to be found. To control whitefly biologically, the use of the parasite *Encarsia* was resurrected from the 1930s (see Section 3.2). A biological control for red spider mite was already being developed in Holland, employing a predatory mite (*Phytoseiulus* spp.) from South America. This biological control of red spider mite was quite complex to operate. The predatory mite was in fact so voracious that chrysanthemum houses needed to be stoked with pest mites after a period to maintain the predator and prevent it dying out. Some growers actually reserved particular houses for rearing the pest in order to keep the predator in the commercial houses supplied, and the situation got even more bizarre when the predator became a problem because it was easily accidentally transported into the pest-mite-rearing houses. The search was on

Fig. 9.1 The aphid *Myzus persicae* on year-round-flowering chrysanthemum. Photograph courtesy of the late Dr. F. Baranyovits.

for a pesticide one had felt would never be needed, a selective pesticide which killed only the biological control agent but left the pest alive! Vapona strips hung in the pest-mite-rearing houses eventually proved effective. The control of thrips posed the problem that no natural enemies that might be manipulated were known. However, several of the important species of thrips go down into the soil around the plants before emerging as adults, and so a soil drench of insecticide could be used to control a large proportion of the thrips population without interfering with biological control agents on the leaves. However, the original problem of the aphids still remained. Rather later on, it was discovered that parasites could be effective in controlling aphids on partially aphid-resistant chrysanthemum varieties, on which the population increase rate of the aphids was somewhat reduced. However, such resistant varieties were not part of the original package which solved the aphid problem. Just at the appropriate time, the carbamate insecticide Pirimicarb, which is biochemically selective for aphids and was originally rejected by industry as not being sufficiently broad spectrum, was marketed. A programme successfully combining biological control of red spider mite and whitefly with insecticidal control of the aphid and thrips was then introduced for chrysanthemums and cucumbers.

9.3 The integrated control concept

If we accept that the main components of pest control which are most likely to be generally applicable are chemical control, biological control, plant resistance and cultural control, then another look at the three examples just quoted will show that the only common feature of the three solutions is that two particular components, chemical control and biological control, were used together in every case. Thus Rachel Carson's 'fork of the roads' position was in no way vindicated; in the event 'insecticide' man and 'biological control' man, we may guess perhaps even to each other's surprise, came together where the two converging, not diverging, roads finally met – the development of integrated control.

Many other examples could be cited, particularly the successful resurrection of biological control in apple orchards of Nova Scotia over a 12-year period. Broad-spectrum pesticides were generally reduced and replaced by a non-persistent plant extract insecticide (rhyania) which allowed the egg parasites of moths (the major pest) to survive. Fungicides were largely replaced by glyodin, which has little effect on the arthropod fauna.

The examples all reflect the original aim of integrated control, defined by Stern, Smith, van den Bosch and Hagen (1959) as 'applied pest control which combines and integrates biological and chemical control. Chemical control is used as necessary and in a manner which is least disruptive to biological control'. The latter sentence suggests that the chemical should be selective between the various life forms which might encounter it in the field, and this aspect is further explored in Section 9.4.5.

In fact, there is some justification for the view that Stern and his co-workers undersold integrated control when they regarded chemical control as being used in a way 'which is least disruptive to biological control'. The examples quoted earlier suggest that integrated control can go much further than this; it can even become predominantly biological control made effective by using insecticides! In Peru, the insecticides which failed had only been introduced in the first place because the beneficial insects were not giving adequate control on their own, and the use of a stomach poison as part of the solution to the problem enabled reintroduced biological control agents to survive. In California, some aphids were still killed with low doses of insecticide and this made the difference between effective and non-effective biological control. In the British chrysanthemum glasshouses it was the introduction of the selective insecticide pirimicarb which made it economically possible to contemplate biological control of red spider mite and whitefly.

The concept of integrated control just described is still the backbone of improved control of individual pest species, and therefore the stages in the practical implementation of integrated control are worth elaborating in a little more detail.

9.4 The procedure of integrated control

9.4.1 Defining the ecosystem

Integrated control is often possible even on a small local scale, particularly in Europe where hedgerows provide a reservoir of fauna untreated by pesticide. However, integrated control is likely to be most effective over a larger area representing a local faunal population, between the patches of which intermixing occurs.

9.4.2 Establishing economic thresholds

This is the important decision of how far a particular pest population can be allowed to grow before insecticide must be applied to control crop loss. Insecticide applications can then be restricted to treatments which are as selective as possible and applied only when absolutely necessary. Stern *et al.* (1959) defined the 'economic threshold' as 'the density at which control measures should be determined to prevent an increasing pest population from reaching the economic injury level'. It is therefore a threshold for action, related by experience and/or experimentation to the 'economic injury level', which is 'the lowest population level that will cause economic damage'. The literature suggests that curves of yield reduction plotted against pest density can take many different forms, and these different forms of curve have often been related to different types of crop. However, to a large extent these different curves merely represent different parts of the generalised relationship illustrated in Fig. 9.2. Low pest infestations may actually be beneficial to yield by stimulating plant growth or by allowing fewer fruits to develop greater size (thus reducing the need for chemical fruit thinners). Some crops (e.g. soya beans) are so leafy that many of the leaves use up assimilate by respiration without contributing greatly by photosynthesis. Defoliation by hand or by insects of a proportion of the leaves on such plants can lead to an increase in yield.

After the phase of stimulation comes the phase of 'compensation'. This is a phase of increasing density which does not reduce yield, since the plant can compensate in some way for damage. Cotton is a good example of a crop in which such compensation occurs. Provided the cotton plant is well fed and watered, a good crop will develop from 12 of the perhaps 90 flower buds that the plant produces. If none of these 90 buds is destroyed by attack of *Lygus* bugs or bollworm, the extra buds will be shed naturally anyway. Crops such as cereals are often sown so densely that adjacent plants are in competition for

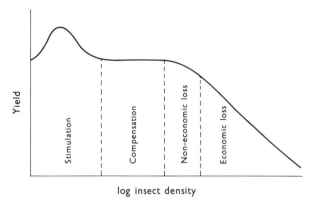

Fig. 9.2 Generalised relationship between crop yield and pest density.

light and nutrients. The result is that cereal crops will give an identical yield per hectare over quite an extensive plateau of plant density. Thus, provided the pest does not destroy cereal plants in patches, adjacent plants will take up the space of killed-out plants, and the larger surviving wheat plants then individually show an increased yield and compensate for the lost plants. Leafiness of crops (referred to above) is often responsible for compensation, especially for leaf damage. Plants give optimum yield at a particular ratio of leaf area to ground area (known as the leaf area index). Thus leaf area damage in excess of this ratio is unlikely to affect yield. One of the interesting aspects of crop improvement by plant breeding is that it tends to be carried out under full insecticide protection; breeders actually seek varieties which stop producing new leaf area after they have achieved the optimum leaf area index. As a result, the more improved that major world crops have become as a result of plant breeding programmes, the more has the capacity for compensation been bred out of the new varieties.

Once damage exceeds the compensating powers of the crop, further increase in pest populations will cause a progressive reduction in yield. However, as long as the damage would not be economic to control, pest levels still lie below the economic injury level. The costs that need to be taken into account are those of pesticide, labour and machinery as well as damage that pesticide application normally causes to the crop (machinery damage, the effects of soil compaction under the tractor wheels and often a measurable check to plant growth resulting from the toxin applied). As long as these costs exceed the damage caused by the pest the cost/potential benefit ratio remains above 1 and the economic injury level has not been reached. Potential benefit is, of course, not a fixed value for a given crop, because it fluctuates with 'environmental' attitudes of the grower concerned, his personal as well as local economic conditions, the state of the market for the crop, the cost of distributing the crop, the investment the crop represents and many other considerations. For example, some growers can obtain a premium for their produce if it bears the scars of pest attack and can therefore be seen to be 'insecticide free'. The introduction of biological control with a pathogen of aphids (Section 4.1) on chrysanthemums had the surprising economic impact that the cut flowers which had not been treated with insecticide could be sold at a premium because of the more attractive soft appearance of the leaves.

In spite of all these variations affecting the relationship of the economic threshold to the economic injury level, thresholds are increasingly used in practice, and it appears that they can be established sufficently accurately to guide the decision 'to spray or not to spray' once the relationship of yield to pest density had been defined for the crop. It is probably also true that thoroughly accurate economic thresholds are not as necessary as was once thought. The real progress that can be made is to educate the farmer away from insurance spraying, so that he reacts only to a challenge from the pest. Whether or not he sometimes then sprays when it is not really necessary is perhaps not as important as that he does reduce the frequency of his spraying. Certainly, it would be much more dangerous for him sometimes not to spray when experience later shows that a yield loss was suffered, as the farmer then loses confidence in the principle of economic thresholds.

9.4.3 Assessing potential natural enemy activity

The crop ecosystem must be sampled to determine whether natural mortality agents are present in sufficient numbers to be worth conserving with selective chemical control, and to determine how frequently the economic threshold is being exceeded.

9.4.4 Augmenting the resistance of the environment

The purpose of augmentation may be to provide natural enemy action where an insufficient one already exists (e.g. heavily sprayed ecosystems, crops with an introduced pest) or to establish a new equilibrium pest population at an artificially low level (Fig. 9.3). Introduced pests can sometimes have their populations equilibrated at a lower level by the importation of natural enemies (especially parasites) to re-establish the stabilising influences of the country of origin from which the pest has escaped. Where parasites have disappeared because a vital alternative host has been lost through monoculture, replacing a single plant species (e.g. planting blackberries near vineyards in California; see Section 3.4.3) may be all that is necessary. The introduction of partial plant resistance to the pest will slow down the rate of pest increase, and the existing natural enemies may well then regulate the pest to a lower equilibrium level. In augmenting the resistance of the environment, cultural controls are also worth considering. Any measure which makes conditions more suitable for natural enemies, such as the provision of refugiae or adult food such as nectar sources, are particularly valuable, as are any measures (such as destruction of crop residues) which may break the life cycle of the pest in the region so that

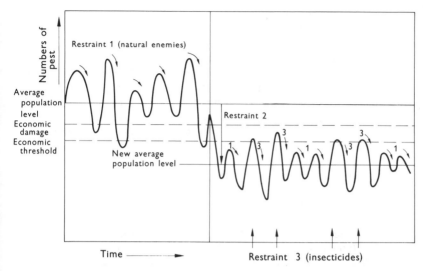

Fig. 9.3 Diagrammatic representation of the role of three restraints in an integrated control programme. Restraint 2 may be an imported natural enemy or a resistant crop variety. From van Emden (1969), courtesy of the Society of Chemical Industry.

numbers in subsequent generations are dependent on immigration from outside.

9.4.5 Developing selective pesticide applications

The biological control potential established or already present in the environment then needs protection from the sprays that are necessary whenever pest populations reach the economic threshold. It would obviously be ideal if we could use chemicals which were inherently selective. However, few inherently selective compounds are likely to be developed. The economic problems of developing insecticides were discussed in Section 1.3.3; very few single crops or pest problems have a large enough usage potential to warrant the development of a specific insecticide tailored to be selective for a particular integrated control solution. Moreover, perhaps too much emphasis has been placed in the past on selectivity between a pest and its natural enemies. There are many pest problems (e.g. low-density pests such as disease vectors) where the pest virtually has to be eliminated. This can usually only be achieved with a pesticide, and the natural enemy specific to the pest might almost just as well be killed by the pesticide as allowed to die of starvation or emigrate because of the disappearance of its prey. In such cases, integrated control involves pesticide selectivity between the pest in question and the natural enemies of other potential pests of the same crop, so that insecticide control of the key pest does not lead to an upsurge of other pest problems.

Apart from the few at least partially selective insecticides which are available, it is often possible to find ways of making a broad-spectrum insecticide selective by the way we use it. When we spray, we will certainly kill both pests and natural enemies; the secret is to make sure that the pesticide application changes the natural enemy to pest ratio in favour of biological control. We may even be able to accept a slightly reduced kill of the pest by our application if we none the less shift the balance in favour of biological control as a result. This exemplifies the principle that it is actually possible to use insecticides to help make biological control more effective. Various sources of selectivity in pesticide applications are discussed below:

a) Inherently selective insecticides

A few pesticides possess selectivity for biochemical reasons. Mention has already been made of the aphicide pirimicarb, a systemic and fumigant carbamate which affects aphids and Diptera, but not ladybirds or aphid parasites (Fig. 9.4). Another widely used insecticide with some selectivity is the organochlorine endosulfan. This compound seems fairly safe to insects in the order Hymenoptera, which includes most of the parasites used in biological control. We should also consider other materials such as insect pathogen sprays (see Chapter 4) and internal growth regulators (see Section 8.5), which tend to be much more selective than traditional insecticides and are therefore more consistent with integrated control. There is increasing research on testing pesticides against natural enemies in the field for particular crop situations. Some selectivity can often be found in such field trials, even with insecticides which are known to be very broad spectrum. For example, in Nigeria the organo-

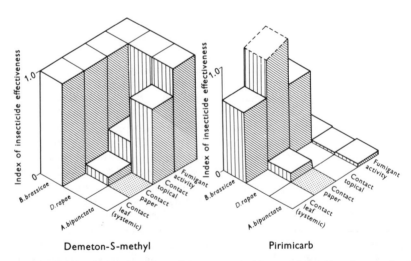

Demeton-S-methyl   Pirimicarb

Fig. 9.4   Comparative toxicities of the contact and fumigant activities of two systemic insecticides to an aphid (*Brevicoryne brassicae*), its main parasite (*Diaeretiella rapae*) and a predator (the ladybird *Adalia bipunctata*). Toxicities are expressed as fractions of the effectiveness of the systemic kill of *B. brassicae* by demeton-S-methyl and pirimicarb. Courtesy Dr. G.D. Dodd.

phosphate methomyl gave good control of a pod borer of cowpeas without affecting its major hymenopterous parasite. Such partial selectivity revealed in field trials is extremely useful, though there may well be other reasons than biochemical selectivity of the insecticide for the observed effect. Selectivity may relate to other factors such as the behaviour of the organisms and how much contact they have with the pesticide.

b) Formulation of spray

Pesticides may be available in different spray formulations, such as emulsions or wettable powders (see Section 1.2). Trying different formulations may reveal differences in selectivity between them, which again may be due to the behaviour of insects in relation to the type of deposit left on the plants, the persistence of different formulations and the contribution of the different additives in the formulation to the toxicity of the deposit. Many insecticides can be sprayed in an encapsulated form to achieve selectivity. Droplets can be coated with polymer; the encapsulated droplets can be sprayed in water or even as formulations which cause the droplets produced from the nozzle to form a capsule on exposure to air on the way to the target. Encapsulation thus converts the contained pesticide into a stomach poison, and natural enemies can move over the leaf safely, since the insecticide only becomes lethal when the capsule is ingested by a leaf-chewing insect.

c) Reduced dose rate

In describing the solution to the problem of resistant lucerne aphids in California (see Section 9.2.2), the crucial concept was that a reduction in dose rate would improve the natural enemy to aphid ratio. Why should this be so? The answer lies in the different type of response (Fig. 9.6a) shown by many carnivores in comparison with many herbivores. The range of insecticide dose spanning low kill to high kill of herbivores is generally larger than that for carnivores. The reason for this is not entirely clear, but may well be connected with the wide range of enzymes, some of which can detoxify pesticides, which herbivores need to cope with the many secondary compounds they encounter in their host plants. Whatever the reason, the result is that the kill of natural enemies decreases faster than the kill for pests as we decrease the insecticide dose; selectivity of the insecticide thus increases until, at a very low dose, it may even be possible still to kill pests without killing any natural enemies at all.

d) Time of application

To obtain selectivity in time requires knowledge of the life history of the natural enemies. There may be times when a high proportion of their population is protected from contact with sprays because they are inside a protective casing (e.g. an insect egg or, for aphid parasites, a mummified aphid) or outside the treated area (e.g. preying on an alternative host in the hedgerow or outside the crop seeking flowers for adult feeding). Recent experiments in Nigeria have suggested that improvements in the ratio of natural enemies to aphids can be obtained by applying the insecticide very early to depress the aphid population before the natural enemies, which arrive in the crops well after the aphids, become abundant. In Pennsylvania apple orchards, populations of spider mites build up in between the two population peaks of the leaf roller *Phyllonorycter blancardella*. By carefully restricting spraying to the early part of June and the latter part of August/beginning of September, insecticides can be used effectively to control the leaf roller without damaging the ladybird populations which are important predators of the mites between these spraying dates. It may even be possible to obtain some selectivity by spraying at a specific time of day. In both cotton and grain legumes, leafhoppers (which normally are difficult targets for spraying since they feed on the undersides of the lower leaves) move up the plant in the evening and sit on the upper sides of the upper leaves. This makes them much easier targets for spraying, and they can be controlled with much less insecticide and much less penetration of the crop, giving the insecticide application considerable selectivity in favour of natural enemies.

e) Application in space

Selectivity in space means treating only a part of the crop to enable natural enemies to survive in the untreated part. A very simple and common way to achieve such selectivity when using systemic insecticides is to apply the toxin as a granule on to the soil rather than as a spray. The roots then take up the poison, but the leaf surface remains safe for natural enemies to move over. In the example quoted earlier of lucerne in California (Section 9.2.2), strip harvesting made it possible for the natural enemies to survive in more recently cut strips whenever the adjacent taller strips required insecticide treatment. This concept of 'band' spraying, where only certain bands of the crop are treated on each occasion insecticide is applied, can be used directly without complications such as strip harvesting. Scale insects in citrus have been controlled by alternating the use of biological and chemical control on adjacent rows on a 2-year schedule; i.e. alternate rows were sprayed with oil emulsion and left clear in alternate years. Insecticide-treated attractive baits may be used to separate the pest from the natural enemies and 'lure it to its doom' (see Section 8.3.2). A very nice combination of a variation on band spraying and the use of attractants has been tried on citrus in Australia against the fruit fly. In several countries, researchers have experimented with restricting spray to the lower half of the tree. Like band spraying, this achieves a fair control of the pest while leaving some of the tree relatively free of pesticide and enabling natural enemies to survive. The Australian workers, however, added an attractant to their insecticide (i.e. they formulated a bait spray), which of course had the result of bringing many flies from the unsprayed upper part of the tree down to the lower part to pick up insecticide there. Perhaps the most ingenious use of restricting pesticide application in space to improve the natural enemy to pest ratio is an example from coffee in the 1950s. One of the most intriguing aspects of this example is that DDT, the widely used broad-spectrum insecticide which had done so much damage to biological control in the 1940s and 1950s, was the agent now used to improve biological control! The main pest of coffee at that time in Kenya, where the technique was used, was the looper caterpillar. DDT was banded around the trunk of each tree in the plantation at the beginning of the season. Whenever looper populations increased to an unacceptable level, the canopy of the tree was sprayed with a natural pyrethrum at a dose to achieve 'knock-down' of insects rather than kill. A large number of insects thus fell out of the trees on to the ground; the caterpillars could only regain the tree by passing over the DDT band and thereby picking up a lethal dose of insecticide, whereas many of the natural enemies, not only of the caterpillars but also of other pests of coffee, regained the tree by flying upwards and thus missed the DDT band altogether.

9.5 The fate of the integrated control concept

9.5.1 Fate of the early examples of integrated control

It is important to realise that, although these early experiments in Peru, California and British glasshouses were highly successful in providing solutions to contemporary insecticide-induced crises, the control systems adopted at that time did not have long-term viability. The Cañete Valley in Peru is now rapidly returning to a monoculture of cotton, strip cutting of lucerne and the use of low-dose insecticide is no longer practised in California and, although many growers still use some biological control in British glasshouses, the package developed for year-round chrysanthemums is not practised widely today. The reason for all this is the advent of the new synthetic pyrethroid insecticides which, at least for a time, are proving effective at controlling the pest problems created by resistance to the earlier insecticides. Thus growers have been able to return to the pest control solution they find preferable, because it is instant and they feel they have some control over it – namely, routine reliance on insecticides. However, insecticide-induced crises will recur and are indeed presently appearing in places around the world, so that work and thinking about integrated control is still needed. In tropical countries (where heavy insecticide use is often uneconomical and pest reproduction rates are very rapid), as more insecticides fail as a result of resistance and as public pressure for reduction in pesticide use escalates, the ideas of integrated control will certainly increase in importance.

9.5.2 Extended concepts of integrated control

Already by the mid-1960s, the original concept of an integration of chemical and biological control had been extended by many people to embrace all suitable pest control methods integrated in a compatible manner (harmonious control). This widened concept of integrated control had the merit that cultural methods were more closely examined for their control potential apart from their effect on natural enemy abundance, but it also brought the danger of 'trying to avoid chemical control' rather than stressing the 'better use of chemicals' on which the original concept had depended. In 1967, at an FAO meeting, 'integrated control' was redefined in terms very similar to harmonious control. By 1970, a new phrase, 'pest management', had been defined as 'the reduction of pest problems by actions selected after the life systems of the pests are understood and the ecological as well as economic consequences of these actions have been predicted, as accurately as possible, to be in the best interests of mankind' (Rabb, 1970).

Pest management is therefore a 'blanket term' for an ecological approach to pest control which emphasises economic and environmental considerations. It is really a definition of what pest control should and might be, rather than a practical protocol of the kind that the first 'integrated control' represented. Pest management equally embraces the multiple approach of integrated control and single component biological control in as much as either may prove the best solution to a particular pest problem. It might be a fair statement that integrated control is likely to prove the most generally applicable pest management solution. Indeed, the principal features of pest management listed by Rabb (1970) and summarised below are in several places very similar to the features of integrated control already mentioned:

(a) The *orientation* is towards entire pest populations rather than to localised ones;
(b) The *proximate objective* is to lower the mean level of abundance of the pest so that fluctuations above the economic threshold are reduced or eliminated;
(c) The *method* or *combination of methods* are chosen to supplement natural control and give the maximum long-term reliability with the cheapest and least objectionable protection;
(d) The *significance* is that alleviation of the problem is general and long-term with minimum harmful side effects;
(e) The *philosophy* is to 'manage' the pest population rather than to eliminate it.

The final stage (up to now) in the development of the jargon has been the definition in 1976 of 'integrated pest management' (Apple and Smith, 1976). Apple and Smith accepted the term 'pest management' in relation to the control of separate groups of crop problems (insects, fungi, nematodes and weeds), but felt that the management of these different categories should be integrated as a total crop protection system. The 'integration' of 'integrated pest management' was therefore of disciplines rather than of methods, as is implicit in the term 'integrated control'. The situation now seems to be that this distinction of integrated pest management has been virtually forgotten, and the terms 'integrated control', 'pest management' and 'integrated pest management' are usually used as synonyms. A recent British text on integrated pest management (IPM) (Burn, Coaker and Jepson, 1987) refers to IPM as 'a control strategy in which a variety of biological, chemical and cultural control measures are combined to give stable long-term pest control'. This definition is given as the simplest form, and the editors go on to say 'we would like readers to form their own opinion as to what IPM is in practice'.

However, although the feature of early integrated control which made it immediately practicable, i.e. the selective use of pesticide, has rather got lost over the last 25 years, another element of that early concept which has gained in acceptance and practical implementation is the use of economic thresholds.

This is perhaps the real key to pest management, because if farmers can be weaned away from routine spraying, then whatever progress research scientists can make in keeping pests below their thresholds by other control measures should be straightforward enough to become integrated with an acceptance by farmers of the principle that insecticides should only be used when necessary.

It is therefore appropriate to discuss the ways in which economic thresholds can be operated before going on to look at how pest management programmes, combining a variety of control methods, can be constructed.

9.6 Monitoring and forecasting

9.6.1 Techniques for establishing economic thresholds

The importance of providing information on the levels of pests which constitute a threat to yield has been discussed previously (Section 9.4.2). A variety of techniques can be used to establish the economic injury level. These are described below, and the economic threshold is then a matter of judgement, giving time for the farmer to take action for the control measure to take effect before the economic injury level itself is reached.

Surveys attempt to correlate yield losses with pest density by recording both over a wide range of the latter. As this often involves surveying a large geographical area where the crop is grown, the very fact that widely differing pest densities can be recorded is itself a warning that correlation with yield may be spurious and confounded with other factors which themselves affect pest density, e.g. longitude, latitude, altitude, soil type or even choice of crop variety typical for an area.

The obvious approach, of infesting plants to known infestation levels, is unfortunately extremely difficult in practice – certainly on a field scale. Many insects are mobile, and tend to move so as to even out initial differences in density. Cages may be needed, but these can change the physiology of the plants grown within them and, through this, pest reproduction rates. With fast-breeding insects such as aphids, it is very hard to relate initial infestation to the population which actually caused damage. Insecticides can, however, be used to create different pest densities experimentally. Many insecticides unfortunately have their own direct effects on crop yield, but this can probably be catered for with suitable controls. Insecticides are certainly useful for checking a rising population at different points in time and seeing which is the latest application which satisfactorily protects yield.

Given the problems of establishing economic injury levels caused by insects, many workers have sought to simulate pest damage by techniques such as using a pair of scissors to create levels of defoliation. Results of simulation experiments usually suggest that the plant can tolerate much more damage than is suggested by equivalent experiments with insects; reasons for this are that insects defoliate bite by bite, and therefore cause far more wounding to the plant than a single cut with a pair of scissors; also insects do not just feed but also add saliva, which itself can severely damage plants.

An interesting technique, which might be termed 'grower bioassay', was used in Kentish apple orchards to get directly at some kind of useful economic threshold. Over several seasons, advisory officers and growers counted pests on the trees and progressively delayed the pest level at which treatments were applied until the grower was convinced he was suffering effects on yield. This technique may not have established scientifically valid economic thresholds, but it certainly gave the perhaps rather more useful information of what was a 'grower worry threshold'!

9.6.2 Monitoring for pest incidence

A direct and commonly used approach is to count the pest on a sufficiently representative number of plants in the crop, a procedure known in many parts of the world as 'scouting'. For example, the published economic threshold for cereal aphids in England is an average of 5 aphids per ear on a sample of 20 tillers. It may not always be necessary to go to the trouble of counting individual pests; for black bean aphids the economic threshold is when 5% of the stems are infested, and other economic thresholds are based on simple measures such as per cent defoliation rather than the number of caterpillars counted.

Traps are increasingly being used for monitoring, since it is usually takes less time for farmers to visit a set number of traps regularly than to walk into their fields and scout individual plants. Here traps emitting sex pheromones (Section 5.1) have proved particularly valuable, providing it can first be shown that there is some correlation between the number of males caught at traps releasing female sex pheromones and the number of eggs subsequently laid on the crop by females. Some examples of the use of pheromone traps for monitoring crop pests in Britain, and the thresholds used, are given in Section 5.1.

Forecasting aims either to predict the need to spray without the grower personally being involved in monitoring or to predict the most effective spray date given the presence of an economic threshold in pest numbers (as in the example of pea moth, Section 5.1). Here climate can be the most important variable, and the presence of an economic threshold of pea moth in the pheromone traps is coupled with the provision of a forecast from the Agricultural Development and Advisory Service of when to spray based on the normal delay between the appearance of males in the traps and the females laying their eggs, plus the delay between the date of oviposition and hatching eggs. This prediction is based on the weather forecast but is updated as actual weather records are taken. Farmers can then refer back to the Advisory Service for a more accurate prediction as the provisional spray date approaches.

The technique of accumulating day degrees has been used for a long time to predict the appearance of pests on the crop after winter. The development and emergence of pests from eggs or pupae is largely determined by temperature, but the temperature has to exceed certain thresholds for any development to take place. In the spring, therefore, it is possible to accumulate the number of degrees of temperature by which the mean temperature for each day exceeds this threshold. When a specified total is reached, the emergence and appearance of the pest can be expected. For example, spraying for the mealybug *Pseudococcus* in apple orchards in Japan is recommended when 220 day degrees above 9°C have accumulated.

Increasingly, computer modelling of plant growth and weather forecasts updated by current weather conditions are being developed to try to produce long-range forecasts from early monitoring data. Some quite long-range predictions relate to the effects of severe winters on pest populations. For example, the annual variation in the incidence of beet yellows virus on sugar beet, vectored by the aphid *Myzus persicae*, can be predicted by calculating the number of days with frost in the previous January to March and how far the following April temperatures deviate from normal. Another example relates to cutworm, where high attack of vegetable crops is correlated with high survival to the third instar. This survival can be modelled from temperature records, corrected for rainfall on the prediction that each 0.1 mm if rain reduces the number of first and second instar caterpillars by 1%.

Egg sampling is used as a basis for a long-range forecast of the incidence of several pest species. For several years the Agricultural Development and Advisory Service has monitored egg populations of wheat bulb fly in fallow fields during August and September. A threshold of 2.5 million eggs per hectare has been established as predicting that spraying will be necessary when the eggs hatch early in the following year, although sometimes high egg counts are not followed by serious damage since the eggs are subject to depredation from ground-dwelling carabid beetles. Since 1970, the Advisory Service has been involved in counting the eggs of the black bean aphid during the winter on selected spindle bushes as a basis for separate long-term forecasts for 16 areas in south England. This has enabled successful predictions about where chemical treatment will be unnecessary or clearly necessary; between 1970 and 1982 only 10% of the total bean area involved received a 'damage possible' forecast, which then required additional shorter term crop monitoring.

Considerable effort has also been devoted to developing forecasts based on evaluating the dispersing populations of some pest species. A major effort centred at Rothamsted Experimental Station in Hertfordshire involves a network of 24 suction traps sampling flying insects at a height of 12.2 m distributed over Britain. Particularly migrant aphids have now been monitored for 20 years, and for some species forecasts have been developed to predict both the likely size of the crop infestation and its timing.

9.7 Pest management packages

Although a synonymy between the words 'integrated control', 'pest management' and 'integrated pest management' seems to have arisen in the minds of many people, the original concept of integrated control was largely directed at individual pest problems. Pest management, on the other hand, is often geared towards the pest complex occurring on a particular crop. The task of a pest manager is therefore usually to develop a package of control measures to deal with the pests of the crop as a complex, rather than in isolation. Three quite different approaches to pest management appear to have evolved:

9.7.1 Synthesis of target-specific controls

With the importance of pests in the cotton crop and the heavy use of insecticides which results, perhaps it is not surprising that work on cotton has led the way in the development of pest management systems. The crop has justified enormous research inputs over a long period, with the result that a great deal is known about non-insecticidal control measures for individual pests. A wide range of measures is available and can be combined in a package appropriate for a particular area. Thus, from the USA, stems the tradition of building pest management packages from individual components. The more important of these are presented in the order in which they have been discussed in the course of this book.

Chemical control is very much based on operating crop scouting and economic thresholds. A chemical frequently used in response to threat is endosulfan, which (as mentioned earlier) does relatively little damage to many of the important parasite species of the crop pests. There have been several attempts to import natural enemies, and one of the main uses of biological control is the release of the parasitic wasp *Trichogramma*, an egg parasite of the bollworm (*Heliothis*). In addition, pheromones can be used in the 'confusion technique' to reduce populations of bollworm. Also, a whole range of resistant plant varieties are available. Early maturing varieties escape late-season bollworm attack, the 'frego bract' characteristic gives high resistance to boll weevil and varieties which are both high in the secondary chemical gossypol as well as devoid of leaf nectaries provide further resistance to bollworm. There is also a wide range of cultural controls. Uniform planting, early termination of irrigation and application of leaf defoliants and desiccants all combine to cause bolls to open up at the same time to allow early harvesting, which reduces bollworm problems, followed by early stalk destruction. Additionally, any adjacent lucerne is strip harvested, as this holds any *Lygus* bugs and restrains them from invading the cotton fields. Insecticide-treated trap crops are sown as a control for boll weevil. Finally, some successes against boll weevil have been achieved with a sterile male technique.

The above type of pest management package has four particular characteristics. First, it is extremely crop-specific; many of the measures relate to particular characteristics of cotton growing and could not easily be transferred to other crops. Second, the control measures are highly target-specific. Each component is designed to reduce levels of a particular pest, and the components have been combined in such a way that the measures against pest A do not interact with attempts to control pest B. Third, the major pest (bollworm) is attacked by a whole range of specific control measures; rather than 'killing two birds with one stone', a veritable basket full of stones is thrown at a single bird! Fourth, the package brings together many years of separate research on the effectiveness of single control measures against individual pests. It is the product of many man-years of work in different research stations and universities, each piece of work not necessarily having been originally envisaged as a contribution towards an integrated pest management system.

9.7.2 Computer design of packages

Much effort towards developing pest management systems has gone into investigating the contribution that might come from the modern ecological tools of life table studies, systems analysis and mathematical modelling. If the role of the various factors which cause changes in insect abundance can be understood and related to predictable events, then a model of the system enables predictions of the consequences of any pest control practices or combinations thereof. Ideally, the system could indeed then be 'managed' to best advantage.

The one obvious problem is that extensive life table data must be collected over several years before a single pest population, let alone that of all the potentially important pests on the crop, can be modelled. Weather and changes in plant growth are important determinants of pest population growth. Elegant computer models of crop growth and the effect of environmental factors on crop growth and pest development have been devised for a number of crops, and computer models are thus already becoming useful for predicting pest incidence in the field. One of the most complete simulation models is that of apple pests developed at Michigan State University in America. It simulates the production and utilisation of assimilate in the whole tree, as well as the growth of leaves, shoots, roots and fruit. The tree model is driven by the environment via the input of weather data and is coupled with developmental models for some eight pest species under the acronym of PETE (Predictive Extension Timing Estimator system). USA orchards make extensive use of natural enemies in the control of mite pests, and four prey-predator models are available to help growers make decisions on maintaining biological control. When appropriate, monitoring assessments from the field are used to synchronise the model to improve the accuracy of the predictions of pest development. Such models can include estimates of control costs and benefits and should allow decision making in relation to control alternatives on the basis of both short- and long-range consequences.

Such modelling is probably the most ecological approach to pest management, and the output of a model will obviously be only as accurate as the information it has been provided with. Although models are already used to predict pest outbreaks, they have as yet contributed rather little to control strategy. The important step towards the 'pest management' goal is not the development of the model in itself, but the early trial of its predictions in the field so that pest management 'output' can be refined and improved in light of experience in the 'real world'; up to now this step has been singularly lacking.

9.7.3 Experimental design of packages

Synthesis and computer design are time-consuming and expensive. A rather different approach is being developed at Reading University in the UK, where the aim is to develop a pest management package in a much shorter time period; the package may then be subject to considerable further refinement and improvement in future years. In contrast with the synthesis approach, and to some extent the computer design approach, the characteristics of the experimental approach are as follows. First, it aims to tackle the whole pest complex from the start, allowing experimental field results to determine the structure of the package rather than previous detailed knowledge about the agronomy of the crop and pest biology and population dynamics. Second, the package seeks to exploit the interactions that can occur between control methods, rather than keeping the inputs highly target-specific and free from interaction. The package design is based on the 'pest management triad' (Fig. 9.5) (van Emden, 1983). The underlying principle is that, if insecticide use is to be reduced, its

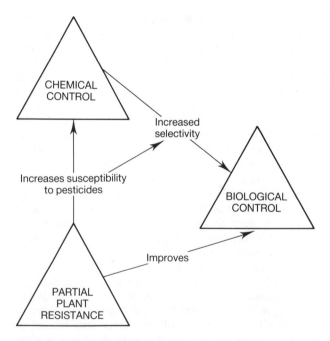

Fig. 9.5 The pest management triad: the interaction of chemical control, biological control and partial plant resistance.

effect must be replaced by another control component, the most generally available of which is likely to be biological control by indigenous natural enemies. Partial plant resistance can often be shown to improve the impact of biological control; there are several reasons for this, probably the most important being that partial plant resistance shifts the ratio between natural enemies and their prey in favour of the natural enemy, particularly where fast-breeding pests are concerned. Partial plant resistance also stresses the pest so that it becomes less tolerant to insecticides. This means that the same kill of the pest can be achieved at a reduced dose, and this in turn improves the selectivity of the application in favour of natural pest enemies (see Section 9.4.5). However, the ability to reduce the dose of insecticide required by growing a partially resistant variety has a much larger effect than discussed earlier in relation to dose reductions. Natural enemies are not affected in the same way as the pest by plant resistance, and their insecticide susceptibility is likely to change relatively little when feeding on prey on a resistant variety. Thus, on a partially resistant variety, the dose-response curves for carnivores and herbivores separate out in such a way as to increase the selectivity window dramatically (Fig. 9.6). The effect of plant resistance on the insecticide tolerance of the pest therefore promotes the impact of biological control still further. The final part

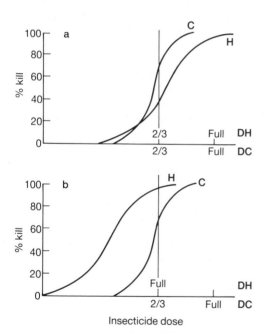

Fig. 9.6 Effect of partial plant resistance on the selectivity of an insecticide. (a) Susceptible variety; (b) resistant variety on which insecticide dose for the herbivore can be reduced by one-third; (c) dose mortality curve for carnivore; (h) dose mortality curve for herbivore; (dc) dose scale for carnivore; (dh) dose scale for herbivore. From van Emden (1987), courtesy of Academic Press.

of the triad is to utilise all the possibilities of achieving greater selectivity of a broad-spectrum insecticide discussed in Section 9.4.5.

The practical implementation of such a system is illustrated in Fig. 9.7. Several criteria can be used to evaluate each stage of the field trials. Since the entire pest complex is involved, harvestable yield is an obvious criterion; so is measuring the intervals between any sprays that become necessary. However, as there is often a trade-off between high efficiency of a single method (e.g. a pesticide) and biological control, other useful criteria include monitoring parasites and predators after the control intervention. In choosing the plant variety on which the package is to be based, one cannot expect total pest spectrum resistance. At this point a decision has to be made to target the plant resistance against either one or more insects which account for the bulk of insecticide use or the pests at a particular stage in crop phenology, e.g. seedling pests, so as to delay the first application of insecticide to the crop for as long as possible. After the selection of a variety, the remainder of the programme involves a sequence of pesticide experiments. Rather than relying on detailed foreknowledge of pest ecology, then, these experimental results can guide us to the best pest management process.

The rationale behind this sequence of experiments is that the number of experimental combinations theoretically possible is enormous and beyond the scope of field experimentation. By taking the experimental programme in defined steps, each step comprises a limited number of treatments for comparison. It is thus a necessary practical feature that the order in which the experiments are done and the decisions made after each experiment eliminate a large number of possibilities which will never be evaluated. Conducting the experiments in a different sequence would almost certainly produce a different final pest management package; the aim is simply to reduce reliance on routine pesticide use in a relatively short time by a programme which is applicable to most crops in most situations.

9.8 Conclusions

Chapter 1 sought to explain that the biggest problem of pesticide use is the development of pesticide-resistant pest populations, and that this is proceeding faster than the development of new approaches to chemical control. Most single alternative methods have problems which prevent them becoming general alternatives to pesticides, particularly in relation to the several pests that often attack one crop. Integrated control and its development to pest management have sought to reduce reliance on pesticides by a multiple control measure approach; this has been most successful where the use of pesticides has augmented other control measures such as biological control. Even so, such ideas have not been readily accepted by farmers in intensive agriculture who can afford high insecticide inputs provided they still have effective insecticides available. Greater acceptance of pest management will almost certainly come from environmental pressures, particularly from the 'anti-insecticide' lobby.

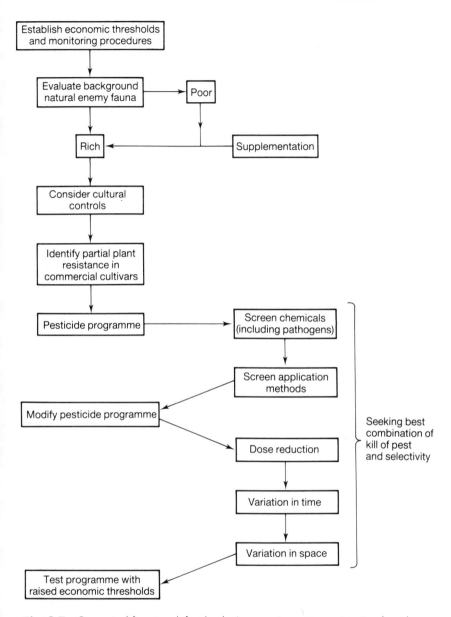

Fig. 9.7 Conceptual framework for developing a pest management system based on experimental variation of crop management.

This is somewhat paradoxical because, by reducing reliance on insecticides, we should be able to preserve the current very useful and flexible arsenal of insecticides from the development of pest resistance to them for far longer. Indeed, without availability of these insecticides, pest management might itself become an impossible goal.

References

Apple, J.L. and Smith, R.F. (eds.) (1976). *Integrated Pest Management*. Plenum, New York.

Birch, M.C. and Haynes, K.F. (1982). *Insect Pheromones*. Studies in Biology no. 147. Edward Arnold, London.

Burn, A.J., Coaker, T.H. and Jepson, P.C. (1987). *Integrated Pest Management*. Academic Press, London.

Carson, R. (1962). *Silent Spring*. Houghton Mifflin, Boston.

DeBach, P. (ed.) (1964). *Biological Control of Insect Pests and Weeds*. Chapman and Hall, London.

Doutt, R.L. (1958). Vice, virtue and the vedalia. *Bulletin of the Entomological Society of America*, **4**, 119–23.

van Emden, H.F. (1965). The role of uncultivated land in the biology of crop pests and beneficial insects. *Scientific Horticulture*, **17**, 121–36.

van Emden, H.F. (1969). Plant resistance to aphids induced by chemicals. *Journal of the Science of Food and Agriculture*, **20**, 385–87.

van Emden, H.F. (1981). Weed plants in the ecology of insect pests. In *Pests, Pathogens and Vegetation* (Thresh, J.M., ed.). Pitman, London. pp. 251–61.

van Emden, H.F. (1987). Cultural methods: the plant. In *Integrated Pest Management* (Burn, A.J., Coaker, T.H. and Jepson, P.C., eds). Academic Press, London. pp. 27–68.

Hassell, M.P. (1976). *The Dynamics of Competition and Predation*. Studies in Biology no. 72. Edward Arnold, London.

Knipling, E.F. (1955). Possibilities of insect control or eradication through the use of sexually sterile males. *Journal of Economic Entomology*, **48**, 459–62.

Lewis, T. (1965a). The effects of shelter on the distribution of insect pests. *Scientific Horticulture*, **17**, 74–84.

Lewis, T. (1965b). The effect of an artificial windbreak on the distribution of aphids in a lettuce crop. *Annals of Applied Biology*, **55**, 513–18.

Matthews, G.A. (1979). *Pesticide Application Methods*. Longman, London.

Painter, R.H. (1951). *Insect Resistance in Crop Plants*. Macmillan, New York.

Rabb, R.L. (1970). Introduction to the conference. In *Concepts of Pest Management* (Rabb, R.L. and Guthrie, F.E., eds). North Carolina State University, Raleigh. pp. 1–5.

Ripper, W.E. (1956). Effect of pesticides on balance of arthropod populations. *Annual Review of Entomology*, **1**, 403–38.

Samways, M.J. (1981). *Biological Control of Pests and Weeds*. Studies in Biology no. 132. Edward Arnold, London.

Solomon, M.E. (1969). *Population Dynamics*. Studies in Biology no. 18. Edward Arnold, London.

Southwood, T.R.E. (1973). The insect/plant relationship – an evolutionary perspective. In *Insect/Plant Relationships* (van Emden, H.F., ed.). Blackwell, Oxford. pp. 3-30.

Southwood, T.R.E. (1975). The dynamics of insect populations. In *Insects, Science and Society* (Pimentel, D., ed.). Academic Press, New York. pp. 151-99.

Stern, V.M., Smith, R.F., van den Bosch, R. and Hagen, K.S. (1959). The integrated control concept. *Hilgardia*, **29**, 81-101.

Index

Index